Biology and Prevention of Sapstain

A conference sponsored by the
Department of Forest Products,
Oregon State University, held May 25, 1997,
Delta Whistler Resort, Whistler,
British Columbia, Canada.

Forest Products Society
2801 Marshall Court
Madison, WI 53705-2295
phone: 608-231-1361
fax: 608-231-2152
www.forestprod.org

The opinions expressed are those of the authors and do not necessarily represent those of Oregon State University or the Forest Products Society.

ISBN 0-935018-99-9

Publication No. 7273

Printed in the United States of America.

9805500

Preface

The development of sapstain on the surface of freshly sawn lumber poses major challenges for those interested in marketing wood on the basis of its bright, clear appearance. For decades, sapstain was controlled by dipping freshly sawn lumber into solutions of sodium pentachlorophenate. This situation changed dramatically in the early 1980s as environmental restrictions encouraged the development of alternative sapstain control chemicals and increased the use of kiln drying. The industry remains in a state of flux and lacks a single, broadly toxic and widely used chemical such as sodium penta. The lack of a magic bullet type of chemical that controls stain on all materials has encouraged a wealth of new research to explore the fundamental mechanisms of fungal discoloration, identify new chemicals, and develop alternative non-chemical protection systems. Sapstain and its control was the subject of a Conference held on May 25, 1997, at the Delta Whistler Resort in Whistler, British Columbia. This Proceedings represents the written papers from that meeting. We realize that preparing the papers entailed an extra effort by the authors and appreciate their willingness to participate in this effort and provide a written record of this conference.

Jeffrey J. Morrell
Department of Forest Products
Oregon State University
Corvallis, Oregon 97331
USA

David J. Dickinson
Senior Lecturer
Imperial College of Science,
Technology & Medicine
Department of Biology
Prince Consort Rd.
London SW7 2BB
United Kingdom

Table of Contents

Wood as a Nutritional Resource for Staining Fungi

Colette Breuil

Abstract

Staining fungi cause wood discoloration, a cosmetic defect that is costly for the forest products industry. These fungi are primary colonizers invading the ray parenchyma and wood cell lumens. They cause only minor weight loss and generally do not affect the structural properties of wood. Most of them grow readily in the sapwood of many tree species; however, they have difficulty in colonizing heartwood. This suggests that nutrients or physical factors that impact fungal growth differ significantly between sapwood and heartwood.

Proteins, carbohydrates, and lipids are the most important food sources for most organisms. These nutrients are present in the sapwood of freshly sawn lumber. To retrieve such nutrients from wood, fungi need to produce extracellular enzymes that hydrolyze these macromolecules into assimilable nitrogen and carbon. In trees, nitrogen concentrations are low. Nitrogen is stored in the sapwood, mainly as proteins. In many wood species, lipid contents are comparable or higher than the levels of soluble sugars and proteins. The lipophilic substances largely consist of triglycerides, fatty and resin acids, sterols, steryl esters, waxes, and fatty alcohols. Staining fungi use lipases to hydrolyze wood triglycerides into fatty acids and glycerol, which are nutrients for fungal growth. The fungi also retrieve nitrogen by producing proteinases and aminopeptidases. In this paper we will describe the nutrients present in the sapwood and heartwood of some commercially important Canadian trees species. We will discuss how the nutritional status of sapwood affects the ability of staining fungi to colonize lumber.

Colette Breuil, Associate Professor, Forest Products Biotechnology, Department of Wood Science, Faculty of Forestry, University of British Columbia, Vancouver, BC, Canada.

Introduction

One of the general challenges facing the solid wood industry is keeping wood, mainly logs and lumber, free from discoloration during processing, storage, and transportation. Stains on wood can be abiotic (e.g., due to iron), but most often stain is due to fungal colonization. Various colors occur, although the most common in sapwood are bluish, bluish-gray to black, or brown (47). Since it is not unusual to find several different fungal species growing in close proximity on a single piece of wood, many fungal species have been associated with stain in living trees, logs, and processed lumber. The fungi most often implicated in sapstain are from the genera *Ceratocystis* and *Ophiostoma* and their anamorphs. Moulds (e.g., *Alternaria*) and black yeasts (e.g., *Aureobasidium pullulans*) are also important, since they can produce a dark penetrating stain (36). In contrast, the green moulds (e.g., *Penicillium, Trichoderma* spp.) discolor wood by forming masses of pigmented asexual spores on the wood surface, which can be brushed or planed off without leaving any residual discoloration (36,45).

Most staining fungi grow primarily on the nutritive substances in the parenchyma cells of the sapwood, and stain the wood by forming granules of melanin within and around the hyphae (4,8,48). Although some sapstaining fungi have been reported to show slight cellulolytic activity in artificial media, no obvious cavities or erosion have been observed in colonized wood (33). Sapstaining fungi seem to lack a complete enzyme system for degrading cellulose and lignin (38). Most of them have little effect on the strength of wood, and the loss of wood dry weight is restricted to a few percent (1% to 4%) after extended colonization (18,35). Similarly, moulds like *Trichoderma* spp., although good cellulase producers in artificial media, do not produce cellulase in wood, and so do not damage wood.

A country-wide survey conducted by Forintek Canada Corporation indicated that staining fungi were isolated from most economically important wood species in Canada. Certain wood species, such as pine, hemlock, and Douglas-fir, are particular prone to stain, while spruce species are less susceptible. Members of the Ophiostomales were prevalent on all woods samples in the Forintek survey (37). These fungi produce ascospores in dark fruiting bodies known as perithecia, and conidia in asexual structures known as synnemata. Pigmented hyphae spread primarily through the sapwood rays, which are rich in nutrients (5).

Wood as a Substrate for Staining Fungi

The microstructure of wood, and nutrients available in wood, affect the growth and survival of colonizing fungi. Wood, composed largely of structural polymers, with minor non-structural components, can be described as consisting of cellulose microfibrils coated with hemicellulose and embedded in lignin. The remaining non-structural wood components are located in the cytoplasm of parenchyma cells, lumens of tracheids and vessels, and in some structures like resin canals. These nonstructural substances can be further categorized into two groups: hydrophilic compounds such as proteins, amino acids, starch, soluble sugars, and minerals; and lipophilic substances, which are also called wood extractives or wood resins.

The lignocellulose that accounts for more than 90 percent of wood dry weight is a polymer mixture rich in carbon but poor in nitrogen (12). The carbon-to-nitrogen ratio of wood varies from about 350:1 to 1,250:1, depending on the tree species, the part of the tree, the location of the tree, and the time of the year (32). In trees, most of the nitrogen is present in organic form. In the fall, trees store considerable amounts of nitrogen as proteins, amino acids, and nucleic acids in the parenchyma cells of wood and bark (27,29,30,34,43,44). In some trees, proteins can account for 75 to 80 percent of the nitrogen in bark and 50 to 60 percent of the nitrogen in the wood. In mature wood only 0.03 to 0.1 percent of the dry weight is nitrogen, while the cambium tissues and newly formed sapwood may contain 10 times this amount. In the sapwood the nitrogen content can be 1.2 to 2.5 times higher than in the heartwood. The low wood nitrogen levels are growth-limiting factors for fungi (31).

In our group we found nitrogen concentrations ranging from 0.045 percent to 0.049 percent in mature wood samples taken at different positions within a 45-year-old lodgepole pine. The nitrogen content decreased slightly from the cambium to the heartwood and was higher in the branches, with a concentration there of 0.066 percent (1). Inorganic nitrogen in the form of ammonia accounted for less than 5 percent and 2 percent of the total nitrogen in lodgepole pine and aspen, respectively (3). Amino acids and low molecular weight proteins were present in mature wood, branches, and sap pressed from green wood chips (1).

Given that staining fungi do not hydrolyze the cellulose in wood, they can use starch and soluble wood sugars as carbon sources (25). The concentrations of these substances vary greatly with the season and among wood species (19,24,27). Starch is stored as grains in the ray and axial parenchyma cells, it increases steadily during the summer and decreases in the fall (23,27,34). The concentrations are higher in sapwood than in heartwood: starch (1.0% to 0.4%, respectively), sucrose (0.4% to 0.2%), glucose (1.5% to 0.5%), and fructose (0.2% to 0.05%). In aspen, we found only very low concentrations of soluble monosaccharides: 0.12 mg/g ovendried (od) wood (42).

Many wood species contain comparable or higher levels of extractives or lipids than soluble sugars and proteins. Wood contains 2 to 6 percent lipophilic extractives, which includes glycerol esters (mono, di, and triglycerides), fatty acids, resin acids, sterols, steryl esters, waxes, fatty alcohols, and volatile compounds (40). However, the values reported in the literature vary from 1.5 percent to more than 10 percent of the dry weight. Extractives have been well characterized in wood species such as *Pinus sylvestris*, *Pinus nigra*, *Picea abies*, and *Populus tremuloides* (19,22,39,46).

Table 1 shows examples of the lipid contents of the sapwood of some wood species. Triglycerides were the most abundant lipid class, representing 36 to 58 percent of the total extractives. Similar results have been reported for other wood species (17). Oleic (18:1) and linoleic (18:2) acids were the most abundant fatty acids residues in the triglycerides, accounting for about 67 to 72 percent of the total residues. In lodgepole pine there was far less free fatty acids (1%) than resin acids (17%). Resin acids are terpenoids. They are present in most softwood species, and absent or present only in trace amounts in hardwood

Table 1.—Lipids in the sapwood of lodgepole pine, white and black spruce, and trembling aspen.

Wood species	Triglycerides		Fatty/resin acids		Waxes/steryl esters	
	(mg/g od)	*(%)*	*(mg/g od)*	*(%)*	*(mg/g od)*	*(%)*
Lodgepole pine	12.8	58	4.1	18	2.2	10
Black spruce	9.8	40	3.7	15	2.2	9.1
Jack pine	13.8	36	8.5	22	3.0	8.0
Aspen	14.5	47	0.3	1	6.2	20

Table 2.—Examples of proteinase and aminopeptidase activities for staining fungi grown in a liquid medium with soy milk as a protein source and in lodgepole pine sapwood. For more extensive results see references 9 and 10.

Fungal strains	Proteinase		Aminopeptidase	
	Medium	Wood	Medium	Wood
	($U\ mL^{-1}$)	(*U/g od*)	($\mu m\ mL^{-1}$)	(*U/g od*)
Ophiostoma piceae E	30	38	3.5	26
Ophiostoma piceae N	28	32	0.3	10
Ophiostoma piceae 212375	30	62	0.2	20
Ophiostoma ainoae	17	47	0.2	14
Alternaria tenuis	16	99	ND	22
Trichoderma harzianum	26	140	1.2	18

Activities were assayed at pH 8.0 and expressed per mL medium or per gram of ovendried (od) wood. In liquid media the activities were measured after 5 days of growth; in wood, after 9 days of growth. Activities were determined using the colorimetric substrates azocoll and L-leucine-para-nitroanilide (1,10).

Table 3.—Percentages of triglycerides and fatty acids in the sapwood of lodgepole pine and trembling aspen after 2 weeks of colonization by different *Ophiostoma* spp.

	Lodgepole pine		Aspen	
	Triglycerides	Fatty acids	Triglycerides	Fatty acids
Control wood	1.3	0.03	1.45	0.03
O. piceae	0.2	0.5	0.27	0.25
O. piliferum	0.3	0.4	0.45	0.3
O. ainoae	0.6	0.2	0.35	0.1

species. Steryl esters are formed from sterols and fatty acids. Waxes and steryl esters were more abundant in aspen than in the softwood species.

Staining Fungi Used Wood Nitrogen for Growth

Organic nitrogen in the form of amino acids, and inorganic nitrogen in the form of ammonia, can be utilized by most fungi (26). For a fungus to utilize exogenous proteins, it has to release extracellular enzymes, proteinases, and peptidases. These enzymes break down proteins into small peptides and amino acids, which can then be assimilated by the fungus (7). We have shown that a wide range of staining fungi produce proteinases and aminopeptidases when growing in protein-supplemented artificial media or in wood (Table 2). The enzymes were synthesized while the fungi were actively growing. In wood, the maximum proteinase activity occurred shortly before or at the beginning of the stationary growth phase, as determined by the ergosterol content (10). The highest proteinase activity in wood was recorded for the green moulds *Trichoderma harzianum*.

We also measured the proteinase activity of *O. piceae* growing on four different sapwood species: lodgepole pine, Douglas-fir, Western hemlock, and aspen (21). The nitrogen contents of the three softwood species were similar at about 0.05 percent, while in aspen the nitrogen was higher at 0.08 percent. The proteinase activity was highest in aspen and lowest in both Douglas-fir and Western hemlock. The major proteinase produced by *O. piceae* was further purified and characterized as a subtilisin-like serine proteinase (2). Degradation of proteins extracted from the xylem tissue of aspen was observed after incubation with the enzyme using SDS-PAGE gels. Polyclonal antibodies were raised against the enzyme, and were used to localize the enzyme release when the fungus grew on lodgepole pine and aspen. Immunogold labeling of the proteinase revealed that the enzyme was secreted into the cell wall and released in a sheath surrounding the hyphae when the fungus grew in wood (21).

Staining Fungi Used Wood Extractives for Growth

When we began work on the ability of sapstaining fungi to degrade and utilize wood lipids as carbon sources for growth, little information on the issue was available. In 1992 Blanchette and his colleagues (6) showed that an albino strain of *O. piliferum* could decrease the total extractives and the esterified fatty acids in southern yellow pine wood chips. In 1994 we also reported the utilization of triglycerides, fatty and resin acids by *Ophiostoma piceae* grown on lodgepole pine (20).

Table 3 shows how triglycerides changed in lodgepole pine and aspen during fungal colonization. The sharp decrease of triglycerides corresponded to a rapid increase in fungal growth. Between 60 to 80 percent of the triglycerides disappeared within the first 2 weeks, while free fatty acids, which were identical to the major fatty acid residues present in wood triglycerides, increased. The results suggested that the fungi secreted extracellular lipases to hydrolyze the triglycerides into glycerol and fatty acids. The glycerol, which is readily assimilated by fungi in liquid media, was probably utilized before the triglyceride-derived fatty acids. At the end of the active growth phase, the different fungi metabolized the fatty acids slowly.

In previous work in liquid media we showed that *O. piceae* was able to use fatty acids (C18:1 and 18:2) as a carbon source (49). The fungal biomass increased with fatty acid concentrations up to 0.5 percent, but did not increase at higher concentrations.

In wood, after fungal colonization, more free fatty acids accumulated in lodgepole pine (about 5 to 7 mg/g ovendried wood after the first week) than in trembling aspen (about 1 to 3 mg/g ovendried), despite the initial triglyceride contents and the rate of triglyceride degradation by the fungi having been similar. This suggested that the fungi assimilated the liberated free fatty acids more efficiently in aspen than in lodgepole pine. Consistent with ergosterol analysis, this could have been related to the higher nitrogen content in aspen supporting better fungal growth than in the softwood species (21).

It is important to note that nutrients can affect fungal pigmentation. *O. piceae* produced a brown pigment when grown in liquid medium with glycerol and asparagine (16). It is likely that a similar induction of hyphal pigmentation takes place in wood. In contrast, fatty acids in solid media, including wood, induce the production of darkly colored perithecia (15).

High concentrations of resin acids are present in some softwood species, and are toxic to many organisms, including fungi (14). Brush et al. (11) and our group (41) showed that *Ophiostoma* species could substantially decrease the level of resin acids in different wood species. Although an association between an increase in fungal biomass and a decrease of resin acids in wood has been shown, such results do not establish whether fungi degrade resin acids and utilize them as a carbon source, or simply modify and detoxify them (28).

Finally, waxes and steryl esters, which accounted for 20 percent of the total wood extractives in trembling aspen sapwood, appeared to be more recalcitrant to fungal degradation (13, and Breuil, unpublished results). We suggested that this lipid class is a poor carbon source for staining fungi.

Conclusions

Literature data and our work have provided direct information on the nutritional physiology of sapstaining fungi, and on their growth in the sapwood of some commercial tree species. To retrieve wood nitrogen, sapstaining fungi and moulds secrete various enzymes. All the staining fungi tested secreted proteinases and aminopeptidases and were able to utilize protein as a source of nitrogen. Similarly, all the *Ophiostoma* species and moulds tested were able to utilize some wood extractives. The fungi secreted lipase to hydrolyze triglycerides, which appeared to be the preferred lipid classes for growth. By further characterizing specific enzyme systems important in the nutrition and pigmentation of staining fungi, it may be possible to manipulate or inactivate fungal physiology and so to disrupt fungal growth or pigmentation.

Literature Cited

1. Abraham, L.D. and C. Breuil. 1993. Organic nitrogen in wood: growth substrates for a sapstain fungus. Doc. No. IRG/WP/10019. International Research Group on Wood Preservation, Stockholm, Sweden. 15p.
2. Abraham, L.D. and C. Breuil. 1996. Isolation and characterization of a subtilisin-like serine proteinase secreted by the sap-staining fungus *Ophiostoma piceae*. Enzyme and Microbial Technology 18:133-140.
3. Abraham, L.D., D.E. Bradshaw, T. Byrne, C. Breuil, and P.I. Morris. 1997. Proteinase as potential targets for new generation anti-sapstain chemicals. Journal forest products (in press).
4. Ballard, R.G., M.A. Walsh, and W.E. Cole. 1982. Blue-stain in xylem of lodgepole pine: a light-microscope study on extent of hyphal distribution. Canadian Journal Botany 60: 2,334-2,341.
5. Ballard, R.G., M.A. Walsh, and W.E. Cole. 1984. The penetration and growth of blue-stain fungi in the sapwood of lodgepole pine attacked by mountain pine beetle. Canadian Journal of Botany 62:1,724-1,729.
6. Blanchette, R.A., R.L. Farrell, T.A. Burnes, P.A Wendler, W. Zimmerman, T.S. Brush, and R.A. Snyder. 1992. Biological control of pitch in pulp and paper production by *Ophiostoma piliferum*. Tappi J. 75:102-106.
7. Breddam, K. 1986. Serine carboxypeptidases. A review. Carlsberg Research Commun 51:83-128.
8. Brisson, A., S. Gharibian, R. Eagen, D.F. Leclerc, and C. Breuil. 1996. Localization and characterization of the melanin granules produced by the sap-staining fungus *Ophiostoma piceae*. Mat. und Org. 30:23-32.
9. Breuil, C. and J. Huang. 1994. Activities and properties of extracellular proteinases produced by staining fungi grown in protein-supplemented liquid media. Enzyme and Microbial Technology 16:1-6.
10. Breuil, C., L.D. Abraham, and C. Yagodnik. 1995. Staining fungi growing in softwood produce proteinases and aminopeptidases. Mat. und Org. 29:15-25.
11. Brush, T.S., R.L. Farrell, and C. Ho. 1994. Biodegradation of wood extractives from southern yellow pine by *Ophiostoma piliferum*. Tappi J. 77:155-159.
12. Carlile, M.J. and S.C. Watkinson. 1994. The fungi. Academic Press, London.
13. Chen, T., Z. Wang, Y. Gao, C. Breuil, and J.V. Hatton. 1995. Wood extractives and pitch problems: analysis and partial removal by biological treatment. Appita 47:463-466.
14. Chung, L.T.K., H.P. Meier, and J.M. Leach. 1979. Can pulpmill effluent toxicity be estimated from chemical analysis? Tappi J. 62:71-74.
15. Dalpé, Y. and P. Neumann. 1976. L'effet des acids gras sur la stimulation des périthèces de *Ceratocystis ips*, *C. minor*, et *C. capillifera*. European Journal Forest Pathology 6:335-342.

16. Eagen, R., A. Brisson, and C. Breuil. 1997. The sap-staining fungus *Ophiostoma piceae* synthesizes different types of melanin in different growth media. Canadian Journal Microbiology (in press).
17. Ekman, R. 1979. Analysis of the non-volatile extractives in Norway spruce sapwood and heartwood. Acta Acad. Abo. Ser. B 39:1-20.
18. Eslyn, W.E. and R.W. Davidson. 1976. Some wood-staining fungi from pulpwood chips. Mem. N.Y. Botany Garden 28:50-57.
19. Fischer, C. and W. Höll. 1992. Food reserves of Scots pine (*Pinus sylvestris*). II. Seasonal changes and radial distribution of carbohydrate and fat reserves in pine wood. Trees 6:147-155.
20. Gao, Y., C. Breuil, and T. Chen. 1994. Utilization of triglycerides, fatty acids and resin acids in lodgepole pine wood by a sapstaining fungus *Ophiostoma piceae*. Mat. und Org. 28:105-118.
21. Gharibian, S., C. Hoffert, L.D. Abraham, and C. Breuil. 1996. Localizing an *Ophiostoma piceae* proteinase in sapwood of four tree species using polyclonal antibodies. New Phytol. 133:673-679.
22. Hafizoglu, H. 1983. Wood extractives of *Pinus sylvestris* L, *Pinus nigra* Arn and *Pinus brutia* Ten with special reference to nonpolar components. Holzforschung 37:321-326.
23. Harms U. and J.J. Sauter. 1992. Changes in content of starch, proteins, fat and sugars in the branchwood of *Betula pendula* Roth during fall. Holzforschung 46:455-461.
24. Hillis, W.E. 1987. Heartwood and tree exudates. Springer-Verlag, Berlin.
25. Hudson, H.J. 1986. Fungal Biology. Edward Arnold Ltd., London.
26. Jennings, D.H. 1989. Some perspectives on nitrogen and phosphorus metabolism in fungi. In: Nitrogen, phosphorus and sulphur utilization by fungi. Boddy, L., R. Marchant, and D.J. Read (eds.) Cambridge University Press, Cambridge pp. 1-31.
27. Kramer, P.J. and T.T. Kozlowski. 1979. Physiology of woody plants. Acad. Press, New York/San Francisco/London.
28. Kutney, J.P., M. Singh, G. Hewitt, P.H. Salisbury, B.R. Worth, J.A. Servizi, D.W. Martens, and R.W. Gordon. 1981. Studies related to biological detoxification of kraft pulp mill effluent. I. The biodegradation of dehydroabietic acid with *Mortierella isabellina*. Canadian Journal Chemistry 59:2,334-2,341.
29. Laidlaw, R.A. and G.A. Smith. 1965. The proteins of the timber of Scots pine (*Pinus sylvestris*). Holzforschung 19:129-143.
30. Langheinrich, U. and R. Tischner. 1991. Vegetative storage proteins in poplar. Plant Physiology 97:1,017-1,025.
31. Levi, M.L. and E.B. Cowling. 1969. Role of nitrogen in wood deterioration. VII Physiological adaptation of wood-decaying and other fungi to substrates deficient in nitrogen. Phytopathology 59:460-468.
32. Merrill, W, and E.B. Cowling. 1966. Role of nitrogen in wood deterioration: amount and distribution of nitrogen in fungi. Phytopathology 56:1,083-1,090.
33. Nilsson, T. 1973. Studies on wood degradation and cellulolytic activity of microfungi. Stu. Forest Suec. 104:2-40.
34. Sauter, J.J. and B. van Cleve. 1994. Storage, mobilization and interrelations of starch, sugars, proteins and fat in the ray storage tissue of poplar trees. Trees 8:297-304.
35. Scheffer, T.C. 1973. Microbiological degradation. In: Wood Deterioration and its Prevention by Preservative Treatments. Vol. 1, pp. 31-106.
36. Seifert, K.A. 1993. Sapstain of commercial lumber by species of *Ophiostoma* and *Ceratocystis*. In: *Ceratocystis* and *Ophiostoma*: Taxonomy, Ecology and Pathogenicity. Wingfield, M.J., K.A. Seifert, and J.J. Webber, (eds.) American Phytopathology Society, Minneapolis, 141-151.
37. Seifert, K.A. and B.T. Grylls. 1991. A survey of the sapstaining fungi of Canada. Unpublished report. Available from Forintek Canada Corp., Sainte Foy, Quebec.
38. Sharpe, P.R. and D.J. Dickinson. 1992. Blue stain in service on wood surface coatings. Part 2: The ability of *Aureobasidium pullulans* to penetrate wood surface coatings. The International Research Group on Wood Preservation. Document no. IRG/WP/1557.
39. Sitholé, B.B., J.L. Sullivan and L.H. Allen. 1992. Identification and quantitation of acetone extractives of wood and bark by ion exchange and capillary GC with spreadsheet program. Holzforschung 46:409-416.
40. Sjöström, E. 1981. Wood chemistry: fundamentals and applications. Academic Press Inc., Orlando, Florida.
41. Wang, Z., T. Chen, Y. Gao, C. Breuil, and Y. Hiratsuka. 1995. Biological degradation of resin acids in wood chips by wood-inhabiting fungi. Applied and Environmental Microbiology 61: 222-225.
42. Wang, Z., R. Leone, and C. Breuil. 1997. Why fungal growth is less effective in aspen heartwood than in aspen sapwood. Mat. und Org. (in press).
43. Wetzel, S. and J.S. Greenwood. 1991a. A survey of seasonal bark proteins in eight temperate hardwoods. Trees 5:153-157.
44. Wetzel, S. and J.S. Greenwood. 1991b. The 32-kilodalton vegetative storage protein of *Salix microstachya* Turz. Plant Physiology 97:771-777.
45. Wilcox, W.W. 1973. Degradation in relation to wood structure. In: Wood deterioration and its prevention by preservative treatments. Nichols, D.D., and W.E. Loos (eds.) Syracuse University Press, Syracuse, N.Y. vol. 1 pp. 107-148.
46. Yildirim, H. and B. Holmbom. 1977. Investigations on the wood extractives of pine species from Turkey. I. Unsaponifiable, nonvolatile, nonpolar components in *Pinus sylvestris* and *Pinus nigra*. Acta Academiae Aboensis, Ser. B. 37(4):1-6.
47. Zabel, R.A. and J.J. Morrell. 1992. Wood Microbiology: Decay and its prevention. Academic Press, San Diego, pp. 326-343.
48. Zheng, Y., J.N.R. Ruddick, and C. Breuil. 1994. Factors affecting the growth of *Ophiostoma piceae* on lodgepole pine heartwood. Mat. und Org. 29:105-117.
49. Zink, P. and D. Fengel. 1990. Studies of the colouring matter of blue-stain fungi. Part 3. Spectroscopic studies on fungal and synthetic melanins. Holzforschung 44:163-168.

Influence of the Nutritional Elements on Pigmentation and Production of Biomass of Bluestain Fungus *Aureobasidium Pullulans*

Anne-Christine Ritschkoff, Marjaana Rättö, and Françoise Thomassin

Abstract

The effect of the carbon source and the amount of nitrogen on the melanization and the production of mycelial mass of bluestain fungus *Aureobasidium pullulans* was studied by using solid state cultivations. The carbon sources used varied from easily soluble sugars to structural polysaccharides existing in lignocellulosic material. The amount of melanin was evaluated by using partial purification and the conventional measurement methods. The production of melanin was clearly dependent on the amount and quality of the carbon source as well as the amount of nitrogen. *A. pullulans* have a high tendency to produce melanin on nitrogen poor media supplemented with easily soluble sugars (e.g., glucose, sucrose, mannose, and xylose). The production was restricted on nitrogen rich media. The production of melanin was totally inhibited on the media containing celluloses or lignin as sole carbon sources.

Introduction

The most critical biological stains are caused by bluestain fungi. Most species causing the discoloration of sapwood belong to the *Ascomycetes*, many to the genera *Ceratocystis* or *Ophiostoma*, or to the *Fungi Imperfecti*. The most striking effect of the invasion of these microorganisms is severe discoloration of wood during storage and the building stage as well as in service. Bluestain fungi are not suggested to decompose wood components by bringing about significant weight losses nor decreased wood strength; although currently it has been shown that some tropical bluestain strains are able to produce both weight losses and decrease wood strength (7,14,15).

The coloring matter of bluestain fungi consists of dark brown or black pigments, melanins. Melanins are found not only in fungi, but also in animals, plants, and bacteria. In the fungal mycelium the melanin pigments are attached to the fungal cell walls (solid medium) or in the extracellular polymers formed around fungal cells (liquid medium). In the *Ascomycetina* and related *Deuteromycetina* the dark brown melanins in the cell walls are generally synthesized from the pentaketide pathway from 1,8-dihydroxynaphtalene (3,5,9,13).

Melanins are not suggested to be essential for growth and development of the fungus, but they enhance the survival and competitive abilities of species in certain environments. The main function of melanins is to enhance the fungal tolerance against environmental and microbial stress. Melanins are considered to be essential for the protection of fungi against UV-IR-radiation, radio waves, dessication, and temperature extremes (3,5,9). Melanins in fungi appear to be important for the resistance to microbial attack, since the melanins inhibit various cell wall digesting enzymes (3). The association of melanins with immune responses has also been noted for plants and invertebrates (2,3).

The aim of the present work is to clarify the influence of the media composition (carbon and nitrogen) on the melanization and the production of mycelial mass and melanization by bluestain fungus, *Aureobasidium pullulans*. The clarification of the environmental and biochemical factors influencing the pigment production is important for understanding the basic mechanisms acting in the pigment formation and for the development of new controlling methods.

Anne-Christine Ritschkoff, VTT Building Technology, Marjaana Rättö, VTT Biotechnology and Food Research, and Françoise Thomassin, Centre Technique du Bois et de l'Ameublement.
This paper was also published as an IRG Document in 1997.

Materials and Methods

The fungal strain of *Aureobasidium pullulans* used in the experiments was obtained from VTT Building Technology collections. The strain was stored and precultivated on solid malt agar (4% malt, 2% agar).

The production of melanins was carried out on solid (2% of agar) culture media containing basic mineral salts (0.1% K_2HPO_4, 0.05% $MgSO_4$, 0.05% KCl, 0.001% $FeSO_4$, and 0.1% yeast extract), carbon source, and nitrogen. The carbon sources used in a concentration of 0.2 percent or 2.0 percent were glucose, sucrose, cellobiose, xylose, mannose, beech xylan (Lenzing), amorphous celluloses (CMC-cellulose and HEC-cellulose), crystalline cellulose (Avicel), and lignin (Indulin). The main nitrogen source used in a concentration of 0.02 percent or 0.2 percent was ammonium sulphate. The solid media containing only 2 percent malt extract was considered as the control. The solid media was covered with a cellophane filter prior the inoculation. The test fungus was inoculated in small malt extract slant. The cultivation was carried out at 22°C for 2 weeks.

The growth speed of *A. pullulans* on different culture media was estimated by measuring the diameter of the fungal colony at the end of the cultivation. The quantity of the fungal biomass was estimated by dry-weight measurements. Prior to the dry-weight measurements, the pigment production was estimated visually and the fungal colony was collected from the cellophane filter covering the solid medium.

The melanin pigments were extracted from the dried mycelium. The extraction was carried out with hot alkali (0.5 M NaOH, 4 hours at 105°C and 20 hours at 20°C) according to Bell & Wheeler (3) and Zink & Fengel (15). The pigments were precipitated from the liquid with 6 M HCl. Acid hydrolysis (pH 1, 18 hours at 105°C) was needed for the degradation of remaining carbohydrates and proteins. The total amount of produced melanin was estimated by dry-weight measurements. The amount of melanin is expressed as a percentage of ovendried fungal biomass.

Results and Discussion

The effect of the carbon source and the amount of the nitrogen on the melanization and the production of mycelial mass of bluestain fungus *Aureobasidium pullulans* was studied by using solid state cultivations. The bluestain fungus, *A. pullulans* is believed to play a central part in initiating the natural breakdown and recycling of plant material, but is also saprophytic on a wide variety of substrates in diverse habitats, from painted surfaces to lymph nodes (10).

The amount and intensity of produced pigments have been observed to depend on the culture conditions (i.e., carbon source and concentration, nitrogen source and concentration, phosphate concentration, amount of trace elements, inoculum source, aeration, initial pH, and temperature) (10). Two different concentrations of carbon source and nitrogen were used in this study. The carbon sources used varied from easily soluble sugars to structural polysaccharides existing in lignocellulosic material. Ammonium sulphate was the chosen nitrogen source according to Deshpande et al. (6), who found the highest level of melanins produced by *A. pullulans* on this nitrogen source. As expected, the fungal growth and production of mycelial mass was most pronounced on simple sugars and malt extract at the concentration of 2 percent. *A. pullulans* was able to grow on simple sugars at a concentration of 0.2 percent and on wood structural polysaccharides when the growth was indicated as the diameter of fungal colony. The production of mycelial mass was, however, only slightly detectable on these carbon sources. Reeslev & Jensen (11) observed that the concentration of yeast extract regulate the fungal growth. In this study the amount of nitrogen did not essentially effect the growth speed or production of mycelial mass of *A. pullulans* (Figs. 1 and 2).

In this study the amount of melanin extracted from *A. pullulans* varied from 0 percent to 10 percent depending on the carbon source used (Table 1). The amount of melanin varies greatly depending the fungal strains used.

Table 1.—The amount of melanins (% of ovendried mycelia) produced by *A. pullulans* on different carbon sources after 2 weeks of cultivation.

Carbon source	Amount of melanin (% of mycelia dry weight)				
	Control	CN++	CN+-	CN-+	CN--
Malt extract	7.1				
Glucose		6.6	6.9	-	-
Sucrose		3.1	4.8	-	-
Cellobiose		4.1	4.4	-	-
Xylose		3.5	4.7	-	-
Mannose		10.2	9.6	-	-
Beech xylan		(+)	+	-	-
CMC-cellulose		-	-	-	-
HEC-cellulose		-	-	-	-
Avicel		-	-	-	-
Lignin		-	-	-	-

CN++ = carbon 2%, nitrogen 0.2%; CN+- = carbon 2%, nitrogen 0.02%; CN-+ = carbon 0.2%, nitrogen 0.2%; CN-- = carbon 0.2%, nitrogen 0.02%.

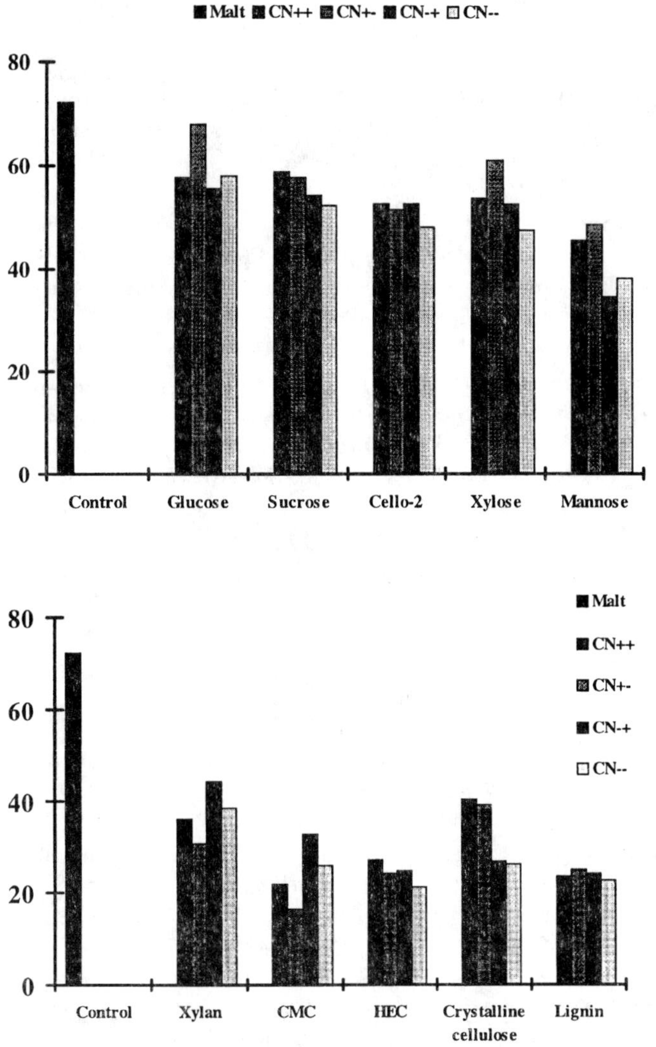

Figure 1.—The ability of *A. pullulans* to grow on different carbon sources: simple sugars (top) and wood carbohydrates (bottom). The values are expressed as diameter of fungal colony (mm). CN++ = carbon 2%, nitrogen 0.2%; CN+- = carbon 2%, nitrogen 0.02%; CN-+ = carbon 0.2%, nitrogen 0.2%; CN-- = carbon 0.2%, nitrogen 0.02%.

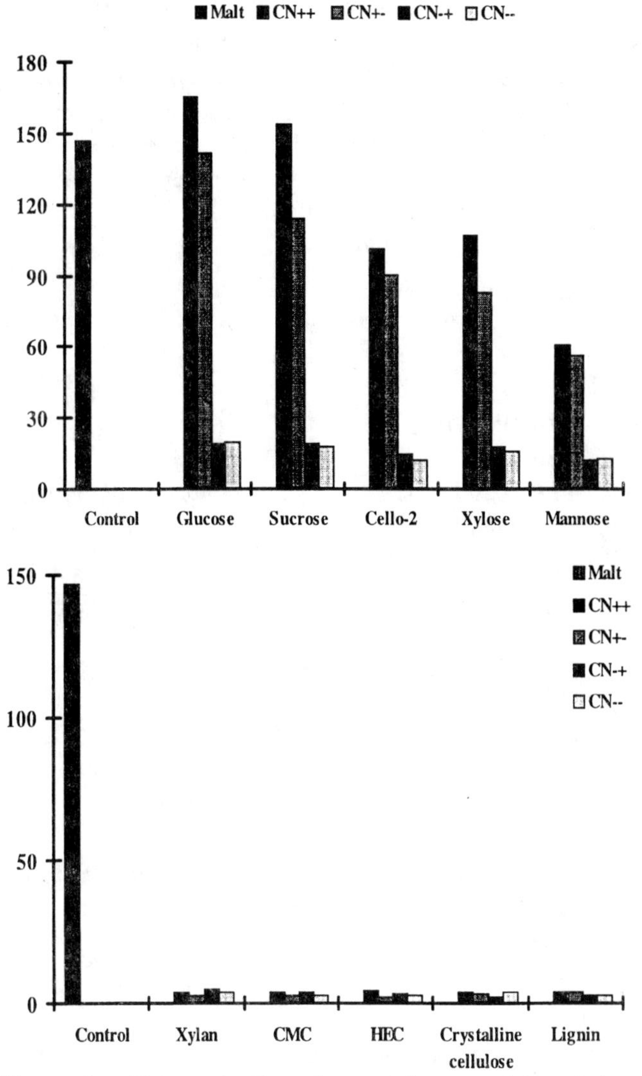

Figure 2.—The production of mycelial mass of *A. pullulans* on different carbon sources: simple sugars (top) and wood carbohydrates (bottom). The values are expressed as mg of ovendried mycelium. CN++ = carbon 2%, nitrogen 0.2%; CN+- = carbon 2%, nitrogen 0.02%; CN-+ = carbon 0.2%, nitrogen 0.2%; CN-- = carbon 0.2%, nitrogen 0.02%.

According to Valmseda et al. (12), the amount of melanin (% of ovendried fungal biomass) produced by different fungi varied between 0.05 percent (*Penicillium citrinum*) and 53.7 percent (*Oidiodendron tenuissimum*). In the studies with *Aspergillus nidulans*, the amount of cell wall bound melanin varied between 16 to 18 percent (4).

The production of melanins of *A. pullulans* was induced on simple sugars and malt extract at concentration of 2 percent. The most pronounced hyphal melanization was observed on mannose (10% of ovendried mycelia), malt extract (7% of ovendried mycelia), and glucose (6.9% of ovendried mycelia). The production of melanins was totally inhibited on carbohydrate poor (0.2%) media or on wood structural polysaccharides. Fairly detectable melanization could be, however, observed on 2 percent hemicellulosic component (Lenzing xylan). Thus, it seems that in addition to simple, soluble carbohydrates, the easily degradable hemicellulosic components act as inductors of hyphal melanization.

The visual detection of the intensity of melanization (results not shown) suggested that the production of melanins was clearly restricted on nitrogen rich media (0.2%). However, by the means of quantitative analysis, there could be observed only slight differences on the melanization between the nitrogen rich and nitrogen poor media. Extraction and partial purification of melanins involve their dissolution in hot alkali and precipitation in acid. Fungal melanins are tightly attached to the polysaccharides and proteins, which were removed by acid hydrolysis (3,8,10,15). Production of extracellular polysaccharide, pullulan, is typical to A. pullulans. It has been shown that this exopolysaccharide production is depending on nitrogen availability so that the rate of production is clearly higher during nitrogen deficiency (11). In this study the lack of significant difference according to nitrogen concentration might be due to the increased production of pullulan and tight attachment of melanins to pullulan on nitrogen poor media and/or to the uncompleted extraction procedure.

The investigation of the factors acting on pigment synthesis and fungal metabolism is important for the development of specific controlling methods. Melanin biosynthesis and production provides an excellent target for highly specific anti-bluestain treatments. It is known, that tricyclazole and some other compound alike, specifically inhibits the biosynthesis of the pentaketide melanin in various fungi. Copper deficiency or copper chelation blocks melanin synthesis, which indicates the involvement of the copper containing laccase or tyrosinase activity in the melanization process (3,6). Banerjee et al. (1) have investigated the proteolytic machinery of bluestain fungi, which has been implicated in their ability to utilize wood proteins for nutrition and growth. The proteinans and other elements acting on nitrogen metabolisms in bluestain fungi could also provide a specific target for novel controlling methods.

Literature Cited

1. Banerjee, S., C. Breuil, and D.L. Brown. 1995. Aminopeptidase activity of staining fungi grown in artificial media. Enzyme and Microbial. technol. 17:347-352.
2. Bell, A.A. 1981. Biochemical mechanisms of desease resistance. Ann. Rev. Plant Physiol. 32:21-81.
3. Bell, A.A. and M.H. Wheeler. 1986. Biosynthesis and functions of fungal melanins. Ann. Rev. Phytopathol. 24:411-451.
4. Bull, A.T. 1970. Chemical composition of wild-type and mutant Aspergillus nidulans cell walls. The nature of polysaccharide and melanin constituents. J. Gen. Microbiol. 63:75-94.
5. Deacon, J.W. 1984. Introduction to modern mycology. Wilkinson, J.F. ed. Blackwell Scientific Publications. Oxford, London. p. 239.
6. Deshpande, M.S., S.G. Bhat, J.C. Thatte, M.A. Kulkarni, and V.B. Rale. 1992. Pigmentation and Phenol Oxidase during Pullulan Production by *Aureobasidium pullulans*. Indian J. Microbiol. 32:95-97.
7. Encinas, O. 1996. Development and significance of attack by Lasiodiplodia theobomae (Pat.) Griff. & Maubl. in Caribbean Pine Wood Species. Acta Universitatis Agriculturae Sueciae. Silvestria 8. Uppsala. p. 43.
8. Gutierrez, A., M.J. Martinez, G. Almendros, F.J. Gonzales-Vila, and A.T. Martinez. 1995. Hyphal-sheat polysaccharides in fungal deterioration. Sci. Total Environm. 167:315-328.
9. Liu, Y.-T., M.-J., Sui, D.-D. Ji, I.-H. Wu, C.-C. Chou, and C.-C. Chen. 1933. Protection from ultraviolet irradiation by melanin of mosquitocidal activity of *Bacillus thuringiensis var. israelensis*. J. Invertebr. Pathology 62:131-136;
10. Pollock, T.J., L. Thorne, and R.W. Armentrout. 1992. Isolation of new Aureobasidium strains that produce high molecular-weight pullulan with reduced pigmentation. Appl. Environm. Microl. 58:877-883.
11. Reeslev, M. and B. Jensen. 1995. Influence of Zn^{2+} and Fe^{3+} on polysaccharide production and mycelium/yeast dimorphism of *Aureobasidium pullulans* in batch cultivation. Appl. Microbiol. Biotechnol. 42:910-915.
12. Valmseda, M., A.T. Martinez, and G. Almendros. 1989. Contribution by pigmented fungi to P-type humic acid formation in two forest soils. Soil Biol. Biochem. 21:23-28.
13. Wheeler, M.H. 1983. Comparisons of fungal melanin biosynthesis in Ascomycetous, Imperfect and Basidiomycetous fungi. Trans. Br. Mycol. Soc. 81:29-36.
14. Zabel, A. and J. Morrell. 1992. Wood microbiology: decay and its prevention. San Diego. Academic Press. 476 p.
15. Zink, P. and D. Fengel. 1988. Studies on the colouring matter of bluestain fungi. Part 1. General characterization and the associated compounds. Holzforschung 42:217-220.
16. Zink, P. and D. Fengel. 1990. Studies on the colouring matter of bluestain fungi. Part 3. Spectroscopic studies on fungal and synthetic melanins. Holzforschung 44:163-168.

Isolation of a Gene from the Melanin Pathway of the Sapstaining Fungi *Ophiostoma Piceae* Using PCR

R. Eagen, J. Kronstad, and C. Breuil

Abstract

To prevent sapstaining fungi from discoloring wood, it is necessary to determine what factors affect the biosynthesis and characteristics of the pigment(s) and to identify the genes involved in the pathway. Using inhibitors and heterologous DNA probes from *Alternaria alternata*, we suggest that melanin, the pigment of *Ophiostoma piceae*, is produced by the dihydroxynaphthalene (DHA) pathway.

Recently, sequences were published for one the enyzmes in the DHN pathway of *Colletotrichum lagenarium*, a cucumber pathogen, and *Magnaporthe grisea*, a pathogen of rice. From this information we synthesized dengenerate oligonucleotides to the conserved regions of the trihydroxynaphthalene (THR1) and tetrahydroxynaphthalene reductase (ThnR) genes. Using these primers and genomic *O. piceae* as template DNA, we obtained a 365 nucleotide PCR product. The deduced amino acid sequence of the product had 85 percent homology to the THR1 of *C. lagenarium* and 80 percent homology to the ThnR of *M. grisea*. This PCR product will be used to screen a genomic library of *O. piceae* in order to isolate the entire gene sequence in the melanin pathway. Complete characterization of the genes involved should facilitate more direct development of anti-stain strategies.

R. Eagen, Forest Products Biotechnology, Faculty of Forestry, J. Kronstad, Biotechnology/Microbiology and Immunology, and C. Breuil, Forest Products Biotechnology, Faculty of Forestry, University of British Columbia, Vancouver, BC, Canada.
This paper was also published as an IRG Document in 1997.

Introduction

British Columbia is one of the world's largest exporter of softwood lumber products. Lumber stored under some conditions while awaiting export is susceptible to stain. The cost of the sapstain problem to the Canadian lumber industry is substantial. Current treatments are either not feasible for all markets or are problematic due to their broad spectrum of action on organisms other than staining fungi. *Ophiostoma piceae* is one of the most commonly isolated sapstaining fungi in Canada (12). It has been previously shown in our laboratory that *O. piceae* produces melanin, the pigment which stains wood (4). Clarifying the pathway of melanin production in this organism may lead to new ways of preventing the stain in lumber that are more environmentally acceptable and economically attractive.

Most ascomycetes produce melanin via the dihydroxynaphthalene (DHN) pathway (17). We suggest that *O. piceae* also utilizes the DHN pathway since several DHN inhibitory compounds block the production of melanin which has also been shown in other fungi (15,18). Also melanin deficient mutants accumulate intermediates of this pathway as do *Berticillium dahliae* and *Thielaviopsis basicola* (15,18). The DHN pathway (Fig. 1) begins with a pentaketide intermediate which is cyclized into 1,3,6,8-tetrahydroxynapthalene followed by two successive iterations of reduction and dehydration to form DHN which undergoes oxidative polymerization into melanin (17). There is some disagreement in the literature on the number of dehydratases and reductases are involved in this pathway.

Recently several of the key enzymes in this pathway have been characterized in other fungi. The polyketide synthase gene (PKS) has been characterized in *Colletotrichum lagenarium* and *Magnaporthe grisea* (6,14).

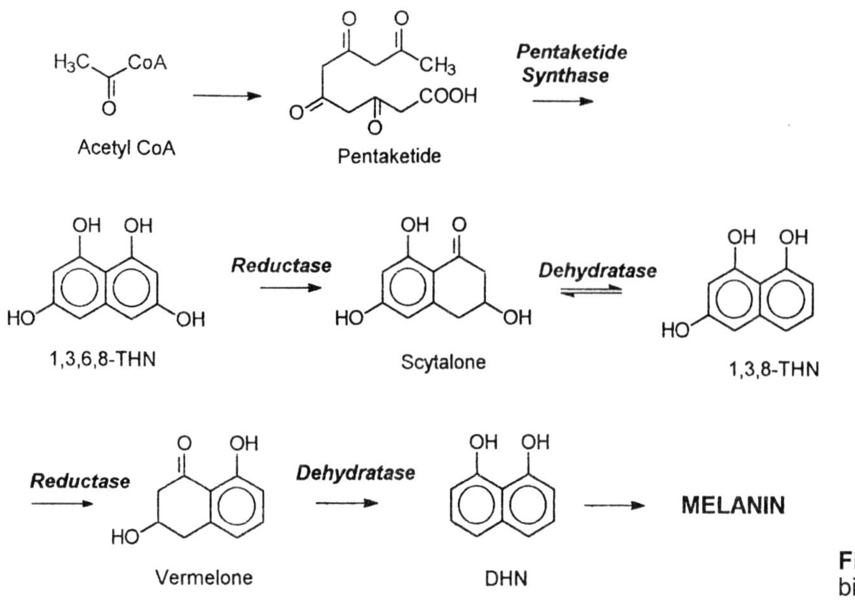

Figure 1.—The DHN pathway of melanin biosynthesis (17). Enzymes are in bold italics.

Scytalone dehydratase (SCD) has been purified and the gene cloned in *C. laganarium*, *M. grisea*, *Cochliobolus miyabeanus*, and *Phaeococcomyces* sp. (5,6,8-10,13). Finally the tetra and tri-hydroxynaphthalenc reductase (THNR) has been studied in both *C. laganarium*, and *M. grisea* (1,2,6,11,16). In the experiment reported here, we used previously published sequences to synthesize degenerate oligonucleotides to the conserved regions of the trihydroxynaphthalene (THR1) and tetrahydroxynaphthalene reductase (ThnR) genes. Using these primers and genomic *O. piceae* DNA as template, we obtained a 365 nucleotide PCR product. The deduced amino acid sequence of the product had 85 percent homology to the THR1 of *C. lagenarium* and 80 percent homology to the ThnR of *M. grisea*. This suggests to us that the *O. piceae* genome contains a similar polyhydroxynaphthalene reductase which may function in the DHN melanin biosynthetic pathway.

Materials and Methods

Fungal Growth

Ophiostoma piceae 387N was obtained from the culture collection of Forintek Canada Corporation. It was cultured in a 2 to 3 percent liquid malt extract in the dark at 23°C and on a rotary shaker at 200 rpm.

DNA Extraction

The fungal mycelia was harvested by centrifugation for 20 minutes at 4,000 rpm in a GS-3 rotor and washed 3 times with 1x PBS. The pellet was then frozen at –70°C and ground into a fine powder with a mortar and pestle and liquid nitrogen. The DNA extraction was carried out according to the method of Zolan and Pukkila (19).

Primer Sequence and PCR Conditions

Degenerate oligonucleotides were synthesized to conserved regions of both the trihydroxynaphalene (THR1) and tetrahydroxynaphthalene reductase (ThnR) genes (11,16). The primers had the following sequences:

T29F 5′GG(CTA)AA(AG)GT(GTC)GCI(CT)TIGT(GTC) ACIGG (TCA)GCIGG3′
T141R 5′(AG)TAIGC(CT)TCIC(GT)IGCIAC(AG)AA (AG) AA(CT)TGICC 3′

where I stands for inosine. PCR was performed with a 50°C annealing temperature for 30 cycles using Taq® Polymerase (Bochringer Mannheim) following the recommended reaction conditions on a Hybaid Omnigene Thermocycler.

DNA Cloning and Sequencing

The PCR products were cloned into the T-tailed plasmid vector pGEM-T (Promega) and transformed into *E. coli* JM109 competent cells and will be referred to as pGT5. Positive transformants were selected on LB/Amp/IPTG/Xgal plates. The PCR inserts were sequenced using T3/T7 sequencing primers, ABI ®PRISM reagents and run on ABI373 automated sequencing apparatus.

Results and Discussion

Previous molecular experiments using heterologous DNA probes from *Alternaria alternata* indicated that THNR was a good candidate for further research (7). Recently two THNR DNA sequences were published allowing comparisons between deduced amino acid sequences (11,16). This comparison revealed several re-

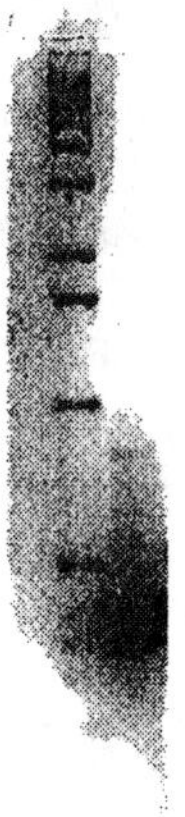

Figure 2.—The 365 bp PCR product from genomic *O. piceae* target DNA and the primers T29F/T141R is indicated in lane 2 of this ethidium bromide stained with 1 percent agarose gel.

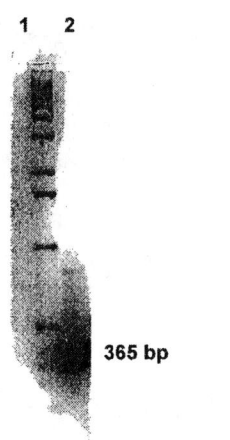

```
                      T29F
          ------------------------------------->
1     GGTAAAGTGGCGCTGGTGACGGGCGCGGGCCGCGGCATTGGCCGCGAGAT     50
      CCATTTCACCGCGACCACTGCCCGCGCCCGGCGCCGTAACCGGCGCTCTA

51    GGCCCTGGAGCTCGGACGCCGCGGCGCCAAGGTCATTGTCAACTATGCCA     100
      CCGGGACCTCGAGCCTGCGGCGCCGCGGTTCCAGTAACAGTTGATACGGT

101   ACAGCGACTCGTCGGCCCAGGAGGTTGTCGATGCCATCAAGGCGGCCGGC     150
      TGTCGCTGAGCAGCCGGGTCCTCCAACAGCTACGGTAGTTCCGCCGGCCG

151   TCCGACGCCGCCGCTATTAAGGCCAACGTCTCCGACGTCGACCAGATTGT     200
      AGGCTGCGGCGGCGATAATTCCGGTTGCAGAGGCTGCAGCTGGTCTAACA

201   CACCCTCTTTGAAAAGACCAAGCAGCAGTGGGGCAAGCTTGACATTGTGT     250
      GTGGGAGAAACTTTTCTGGTTCGTCGTCACCCCGTTCGAACTGTAACACA

251   GCTCCAACTCGGGCGTCGTCAGCTTTGGCCATGTCAAGGATGTCACGCCC     300
      CGAGGTTGAGCCCGCAGCAGTCGAAACCGGTACAGTTCCTACAGTGCGGG
                                                    T141R
                                          <-------------
301   GAGGAGTTTGACCGCGTCTTCTCCGTCAACACCCGCGGCCAGTTCTTCGT     350
      CTCCTCAAACTGGCGCAGAAGAGGCAGTTGTGGGCGCCGGTCAAGAAGCA
      ---------------
351   CGCCCGCGAAGCCTA
      GCGGGCGCTTCGGAT
```

Figure 3.—Nucleotide sequence of pGT5. The arrows indicate the positions of the PCR primers used to recover this fragment.

gions with a very high degree of amino acid residue conservation. With a codon usage table from *O. ulmi* (3), we were able to work backwards from the protein sequence and design four short oligonucleotides that had minimal nucleotide ambiguity. In highly ambiguous positions we substituted inosine.

These oligonucleotides were used as primers in a PCR reaction with *O. piceae* DNA as a template. Knowing the locations of the primers within the original THNR sequence allows us to predict the expected size of the PCR product. Two primers, T29F and T141R (the locations are indicated in Fig. 3), gave a major band of the predicted size of 365 base pairs (bp). This band is indicated on an ethidium bromide stained 0.7 percent agarose gel in Figure 2. The 365 bp PCR product was isolated, purified, and subcloned into a T-tailed vector pGEM. This plasmid, now called pGT5, was transformed into competent *E. coli* JM109 cells.

```
      29
Thr1  GKVALVTGAGRGIGREMAMELGRRGAKVIVNYANSAETAEEVVQAIKKSG
ThnR  ***********************C************T*S*****A****N*
Op    ******************L****************DSS*Q***D***AA*

      79
      SDAASIKANVSDVDQIVKMFGEAKQIWGRLDIVCSNSGVVSFGHVKDVTP
      ****CV****GV*ED**R**E**VK*F*K*********************
      ****A************TL*EKT**Q**K*********************

      129
      EEFDRVFAINTRGQFFVARE
      *******T************
      *******SV***********
```

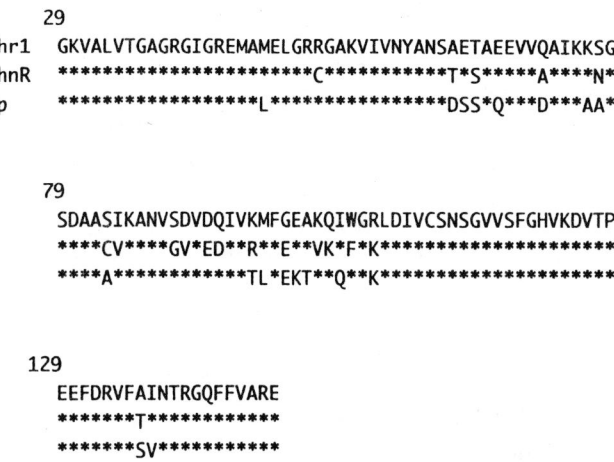

Figure 4.—Alignment of the putative amino acid sequence of the *O. piceae* (Op) with *C. lagenarium* THR1 and *M. grisea* ThnR.

The plasmid pGT5 was then sequenced in both directions using commercially available primers. The sequence is shown in Figure 3. Figure 4 is the putative amino acid sequence of pGT5 aligned with the sequences of THR1 from *C. lagenarium* and ThnR from *M. grisea* (11,16). The asterisks indicate complete sequence conservation. The numbers correspond to the amino acid positions within the gene sequence of THR1 from *C. lagenarium*. The *O. piceae* PCR sequence shares 85 percent homology to THR1 of *C. lagenarium* and 80 percent homology to ThnR of *M. grisea*.

Conclusions

The recovery of this PCR product and its high homology to other fungal THNR's may indicate the presence of this gene in *O. piceae* and account for the alteration of pigmentation when DHN inhibitors are used.

Acknowledgments

This work was supported by a Science Council of British Columbia/GREAT Scholarship co-sponsored by Tony Byrne and Dr. Dave Plackett of Forintek Canada Corporation.

Literature Cited

1. Andersson, A., D. Jordan, G. Schneider, B. Valent, and Y. Lindqvist. 1996. Crystallization and preliminary x-ray diffraction study of 1,3,8-trihydroxynaphthalene reductase from *Magnaporthe grisea*. Proteins 24:525-527.
2. Andersson, A., D. Jordan, G. Schneider, and Y. Lindqvist. 1997. A flexible lid controls access to the active site in 1,3,8-trihydroxynaphthalene reductase. F EBS Letters 400:173-176.
3. Bowden, C.G., W.E. Hintz, R. Jeng, M. Hubbes, and P.A. Horgen. 1994. Isolation and characterization of the cerato-ulmin toxin gene of the Dutch elm disease pathogen, *Ophiosionia ulmi*. Current Genetics 25:323-329.
4. Brisson, A., S. Gharibian, R. Eagen, D. Leclerc, and C. Breuil. 1996. Localization and characterization of the melanin granules produced by the sap-staining fungus *Ophiostoma piceae*. Mat. und Org. 30(1):23-32.
5. Butler, M.J., G. Lazarovits, V.J. Higgins, and M. Lachance. 1988. Partial purification and characterization of a dehydratase associated with the pentaketide melanogenesis pathway of *Phaeococcomyces* sp. and other fungi. Exp. Mycol. 12:367-376.
6. Chumley, F.G. and B. Valent. 1990. Genetic analysis of melanin deficient, nonpathogenic mutants of *Magnoporthe grisea*. Mol. Plant, Microbe, Interact. 3(3):135-143.
7. Kimura, N. and T. Tsuge. 1993. Gene cluster involved in melanin biosynthesis of the filamentous *Alternaria alternata*. J. Bact. 175(14):4,427-4,435.
8. Kubo, Y., Y. Takano, N. Endo, N. Yasuda, S. Tajima, and I. Furusawa. 1996. Cloning and structural analysis of the melanin biosynthesis gene SCD1 encoding scytalone dehydratase in *Colletotrichum lagnarium*. App. Env. Micro. 62(12):4,340-4,344.
9. Lundqvist, T., P. Weber, C.N. Hodge, E.H. Braswell, J. Rice, and J. Pierce. 1993. Preliminary crystallographic studies on scytalone dehydratase from *Magnaporthe grisea*. J. Mol. Biol. 232:999-1,002.
10. Lundqvist, T., J. Rice, C.N. Hodge, G. Basarab, J. Pierce, and Y. Lingqvist. 1994. Crystal structure of scytalone dehydratase—a disease determinant of the rice pathogen *Magnaporthe grisea*. Structure 2:937-944.
11. Perpetua, N.S., Y. Kubo, Y. Takano, and I. Furusawa. 1996. Cloning and characterization of a melanin biosynthetic THR1 reductase gene essential for appressorial penetration of *Colletotrichum lagenarium*. Mol. Plant. Microbe. Interact. 9(5):323-329.
12. Seifert, K.A. and B.T. Grylls. 1990. A survey of the sapstaining fungi of Canada. Forintek Canada Corp. Report. Ottawa, Ont. Canada.
13. Tajima, S., Y. Kubo, and J. Shishiyama. 1989. Purification of a melanin biosynthetic enzyme converting scytalone to 1,3,8-trihydroxynaphthalene from *Cochliobolus miyabeamus*. Exp. Mycol. 13:69-76.
14. Takano, Y., Y. Kubo, K. Shimizu, K. Mise, T. Okuno, and I. Furusawa. 1995. Structural analysis of PKS1, a polyketide synthase gene involved in melanin biosynthesis in *Colletotrichum lagenarium*. Mol. Gen. Genet. 249:162-167.
15. Tokousbalides, M.C. and H.D. Sisler. 1979 Site of inhibition by tricycclazole in the melanin biosynthetic pathway of *Verticillium dahliae*. Pest. Biochem. Physio. 11:64-73.
16. Vidal-Cros, A., F. Viviani, G. Labesse, M. Boccara, and M. Gaugry. 1994. Polyhydroxy-napthalene reductase involved in melanin biosynthesis in *Magnaporthe grisea*. Eur. J. Biochem. 219:985-992.
17. Wheeler, M.H. and A.A. Bell. 1988. Melanins and their importance in pathogenic fungi. In: Current topics in medical mycology. McGinnis, M.R. (ed.) vol. 2. Springer-Veralg, NY.
18. Wheeler, M. and R. Stipanovic. 1979. Melanin biosyntheses in *Thielaviopsis basicola*. Exp. Mycol. 3:340-350.
19. Zolan, M. and P. Pukkola. 1986. Inheritance of DNA methylation in *Coprinus cinereus*. Mol. Cell. Bio. 6(1):195-200.

Prevention of Sapstain in Logs Using Water Barriers

Robin Wakeling, Diahanna Eden, Colleen Chittenden, Ben Carpenter, Ian Dorset, Jon Wakeman, and Richard Kuluz

Abstract

Placing logs in ponds or under water sprinklers to keep moisture levels high enough to prevent oxygen tensions rising above inhibitory levels is a proven method of preventing sapstain. The objective of this experiment was to determine if water barriers significantly reduce the rate of sapstain penetration into logs by maintaining moisture contents, and consequently oxygen tensions, at inhibitory levels.

Debarked radiata pine (*Pinus radiata* D. Don) log billets (20 to 25 cm diameter by 1 meter long) were treated with several different formulations of water barrier, water barrier plus a commercial anti-sapstain formulation, and the commercial formulation used alone. Log billets were subjected to two storage regimes: 25°C and 75 percent relative humidity and ambient temperature and humidity in a pole barn, for periods up to 16 weeks. Billets stored at 25°C were artificially inoculated with sapstain fungi. The extent of internal sapstain and external fungal degrade was assessed at 4, 8, and 16 weeks.

The results showed that a water barrier applied as a layer approximately 0.5 to 1 mm thick, to radiata pine stored under the conditions employed, gave significantly (5% level of probability) better protection from sapstain compared to the commercial anti-sapstain formulation. There was a strong positive correlation between the rate of moisture loss and the rate of sapstain development which suggested that moisture retention was the main mechanism of action of water barrier coatings. While the water barrier on it's own can give superior protection from sapstain compared to the commercial standard used, the water barrier plus the commercial standard performed significantly (5% level of probability) better than the water barrier alone.

Robin Wakeling, Diahanna Eden, Colleen Chittenden, and Ben Carpenter, New Zealand Forest Research Institute, Rotorua, New Zealand and Ian Dorset, Jon Wakeman, and Richard Kuluz, Chemcolour, (NZ) Ltd., Auckland, New Zealand.

Introduction

Placing logs in ponds or under water sprinklers to keep moisture levels high enough to prevent oxygen tensions rising above inhibitory levels is a proven method for controlling sapstain. The objective of this experiment is to determine if a water barrier significantly reduces the rate of sapstain fungi penetrating into logs by maintaining moisture contents, and consequently oxygen tensions, at inhibitory levels.

Materials and Methods

Treatments

The following treatments were used:

1. 20 percent emulsified water barrier (EWB20),
2. Commercial formulation (CC375) containing oxine copper (0.375% w/v) plus carbendazim (0.375% w/v),
3. EWB20 plus CC375,
4. 40 percent emulsified water barrier (EWB40),
5. EWB40 plus CC375,
6. Hot water barrier (HWB) plus CC375,
7. Untreated control.

Log Billets

Pruned radiata pine logs 20 to 25 cm in diameter were felled and debarked using a commercial maul debarker. They were sawn into 1-meter-long billets using a chainsaw.

Anti-Sapstain Treatment

A galvanized nail (200 mm) was hammered to one third of it's length into the center of the cross-cut ends of billets to provide a purchase for a rope loop used to maneuver the billets. Billets were given a 5-second dip in a 200 liter tank containing 100 liters of treatment solution and placed on wooden bearers until dripping had ceased. Billets dipped in CC375 were end-sealed after dipping by coating the cross-cut ends with Mulseal, a bitumen based waterproofing material. Billets treated with a water barrier plus CC375 were dipped in the CC375 first, end-sealed, and then dipped in the water barrier. Billets treated with water barrier alone were end-sealed before dipping.

Storage Regimes

Warm room.—Billets were placed in an alkyd painted iron frame housed in a room held at 25°C and 75 percent relative humidity with an air exchange of approximately 3.2 m^3/min^{-1}. Billets of each treatment were kept separate using iron girder uprights bolted to the bearers.

Pole barn.—Billets of each treatment were stacked in a pyramid in a barn which was open at both ends and had slatted sides. The barn was located in an exposed situation and was ventilated. The trial was conducted during the summer.

Assessment of Sapstain

Extent of sapstain was assessed for 5 replicate billets at 4, 8, and 16 weeks after treatment. Billets were assessed for surface coverage of fungi using the following rating scale:

Rating	*% surface coverage of fungal degradation*
0	zero
1	1 to 5
2	6 to 25
3	26 to 50
4	51 to 75
5	76 to 100

Billets were then cut into 5 biscuits, and the depth of sapstain penetration was then determined for 5 distinct wedges of sapstain that represented the 5 deepest zones. Surface cover of sapstain on the cross-cut surface was also rated using the above scale. A mean sapstain penetration value was calculated for each billet from a total of 20 values (5 wedges by 4 biscuits) and for each treatment using the 5 replicate billet means. Treatment means were similarly calculated for percent cover of external surface fungal degrade and percent cover of internal sapstain.

Weight Loss

Five billets of each treatment were weighed immediately after treatment and again after 8 weeks to determine weight loss.

Statistical Analysis

The most important indicator of the performance of an anti-sapstain chemical is the extent of internal sapstain. Depth of penetration or percentage cover of sapstain on cross-cut biscuits both give a reasonable indication of the degree of internal sapstain and in general differences between treatments are similar for both types of data. For the purposes of this study the depth of penetration data is used for statistical analysis. Using a SAS statistical analysis program an analysis of variance followed by a least significant difference test was performed. Due to non-homogeneity of the data an arcsin transformation was carried out before the analysis of variance.

Experimental

Two separate trials were carried out. In trial one, 10 replicate billets were dipped in treatments 1 through 3 and in trial two 10 billets were dipped in treatments 4 through 6. Ten untreated control billets were included for each trial. For trial two debarked logs were treated 8 hours after felling and 30 hours after felling for trial one (due to a complication). For trial one billets were wrapped in plastic before treatment. After treatment, billets were held in either a warm room or in a pole barn at ambient temperatures for storage until assessments at either 4, 8, or 16 weeks (Table 1).

Log billets stored in the warm room were artificially inoculated with *Ophiostoma piceae* H & P Syd and *Diplodia pinea* (Desm.) Kickx., the two predominant sapstain fungi of radiata pine in New Zealand. An agar plate of a 14-day-old culture of the two fungi was blended with 500 mls of water and then diluted to give an aqueous

Table 1.—Assessment and storage regime used for log billets.

Trial no.	Treatment[a]	Assessment time	Storage regime
		(weeks)	
1	EWB20	4 and 8	warm room
1	EWB20	4, 8, and 16	ambient
1	CC375	4 and 8	warm room
1	CC375	4, 8, and 16	ambient
1	EWB20 + CC375	4 and 8	warm room
1	EWB20 + CC375	4, 8, and 16	ambient
2	EWB40	8 and 16	ambient
2	EWB40 + CC375	8 and 16	ambient
2	HWB + CC375	4 and 8	warm room
1	Control	8 and 16	ambient
1 and 2	Control	4 and 8	warm room

[a] CC375 - commercial anti-sapstain formulation; EWB20 - 20% emulsified water barrier; EWB40 - 40% emulsified water barrier; HWB - hot water barrier; Control - untreated.

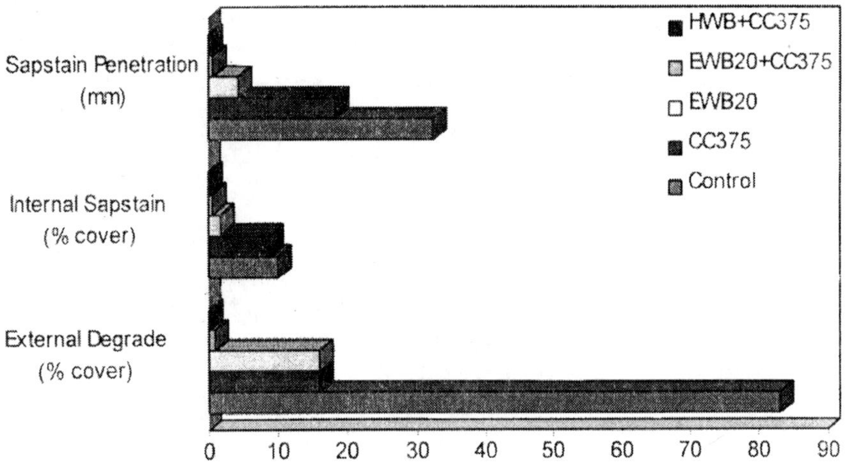

Figure 1.—Fungal degrade of log billets after 4 weeks at 25°C and 75 percent relative humidity.

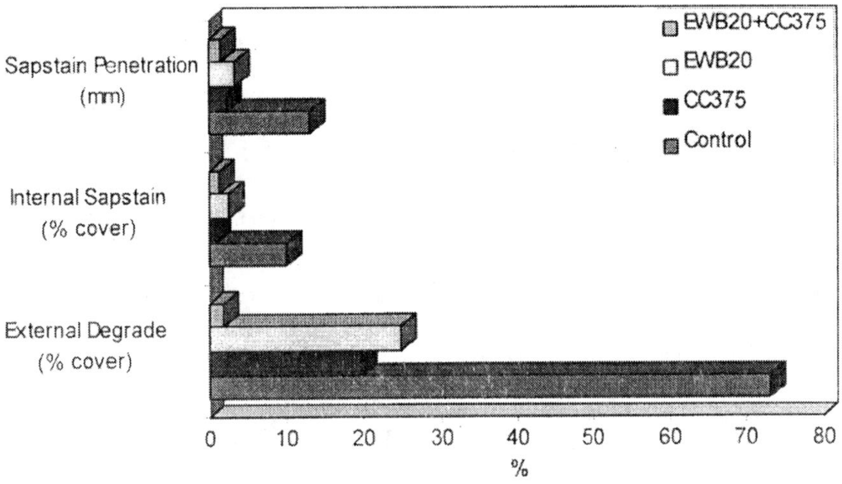

Figure 2.—Fungal degrade of log billets after 4 weeks at ambient temperature.

Table 2.—Effect of treatments[a] on mean sapstain penetration and significant differences.

	Sapstain penetration (mm)				
	4 weeks		8 weeks		16 weeks
Treatment[b]	25°	Ambient	25°	Ambient	Ambient
EWB20	4 (A)	3 (A)	12 (A)	4 (B)	58 (D)
CC375	18 (B)	2.35 (A)	19 (B)	2 (A/B)	23 (C)
EWB20 + CC375	1 (A)	2 (A)	27 (B)	1 (A/B)	30 (C)
EWB40		–		2.5 (A/B)	7.3 (B)
EWB40 + CC375		–		0.3 (A)	0.6 (A)
HWB + CC375	0 (A)	–	0 (A)		–
Control	32 (C)	13 (B)	84 (C)	23 (C)	72 (E)

[a] Treatments with the same letter in a column do not differ significantly (5% level of probability) from each other. A performed significantly better than B which performed better than C, etc.

[b] CC375 - commercial anti-sapstain formulation; EWB20 - 20% emulsified water barrier; EWB40 - 40% emulsified water barrier; HWB - hot water barrier; Control - untreated.

suspension of spores and hyphae of approximately 1 by 10^4ml. The suspension consisted of hyphae of *D. pinea* and *O. piceae* and conidia of the *Graphium* anamorph of *O. piceae*. The top and sides of the billet stack of each treatment group (two billets wide) was lightly sprayed.

Results

Fungal Degrade

Figure 1 shows the fungal degrade of log billets stored at 25°C and 75 percent relative humidity for 4 weeks, and Figure 2 shows 4 weeks of storage at ambient temperatures. For log billets stored at 25°C and 75 percent relative humidity for 8 weeks, the degree of external fungal degrade and internal sapstain was considerably less for billets treated with CC375 plus HWB compared to billets treated with CC375 alone (Table 2).

The untreated control had a mean external cover of fungal degrade of 58 percent, compared to 11 percent for CC375 (trial one) and 2.8 percent for HWB plus CC375. The internal mean cover (on cross-cut discs) was 48

Table 3.—Mean fungal degrade of logs.

Trial no.	Treatment[a]	Exposure period	Exposure regime	External fungal degrade		Internal sapstain			
						% Cover		Penetration (mm)	
				Mean	SD	Mean	SD	Mean	SD
1	Control	4	Warm	83	11	10	3	32	15
1	CC375	4	Warm	16	0	9	7	18	13
1	EWB20	4	Warm	16	0	2	1	4	3
1	CC375 + EWB20	4	Warm	1	0	1	1	1	1
1	Control	8	Warm	59	32	48	35	84	12
1	CC375	8	Warm	11	7	10	4	19	21
1	EWB20	8	Warm	20	10	4	2	12	5
1	CC375 + EWB20	8	Warm	3	0	6	8	27	37
1	Control	4	Ambient	73	14	10	7	13	8
1	CC375	4	Ambient	19.6	17.61	0.98	1.1	2.35	2.2
1	EWB20	4	Ambient	25	12	3	1	3	1
1	CC375 + EWB20	4	Ambient	2	1	1	0	2	1
1	Control	8	Ambient	83	11	13	18	23	18
1	CC375	8	Ambient	8	8	0	1	2	2
1	EWB20	8	Ambient	39	24	3	1	4	3
1	CC375 + EWB20	8	Ambient	8	7	5	6	1	1
1	Control	16	Ambient	76	14	35	23	72	21
1	CC375	16	Ambient	18	13	8	5	23	15
1	EWB20	16	Ambient	83	1	21	10	58	7
1	CC375 + EWB20	16	Ambient	64	34	11	4	30	14
2	CC375	4	Warm	18	13	2.3	1.3	3.45	2.91
2	HWB + CC375	4	Warm	0	0	0	0	0	0
2	Control	8	Warm	58	21	56	8.1	77.72	5.5
2	CC375	8	Warm	39	17	3.5	1.5	24.9	6.9
2	HWB + CC375	8	Warm	6.8	8	0.2	0.3	0.35	0.65
2	EWB40	8	Ambient	9.9	7.7	1.2	1.4	2.5	1.9
2	EWB40 + CC375	8	Ambient	6.8	8	0.3	0.4	0.3	0.6
2	EWB40	16	Ambient	16	0	2.6	1.6	7.3	5.6
2	EWB40 + CC375	16	Ambient	2.4	1.3	0.5	0.7	0.6	1.2

[a] CC375 - commercial anti-sapstain formulation; EWB20 - 20% emulsified water barrier; EWB40 - 40% emulsified water barrier; HWB - hot water barrier; Control - untreated.

percent for the control (trial one), 10 percent for CC375, and 0.2 percent for HWB plus CC375. The mean depth of penetration was 84 mm for the control, 19 mm for CC375, and 0.35 mm for HWB plus CC375.

After 16 weeks of storage at ambient temperature and humidity, logs treated with EWB40 plus CC375 had an average depth of sapstain penetration of 0.6 mm compared to 23.26 mm for CC375 alone.

Comparison of the extent of sapstain in billets treated with water barrier alone (EWB40 and HWB) and those treated with water barrier and CC375 clearly shows that the best level of protection from sapstain is achieved when the two are used together.

There was some variability of results between trials for CC375 which was included in both, for example surface cover of fungal degrade values were 11 percent for trial

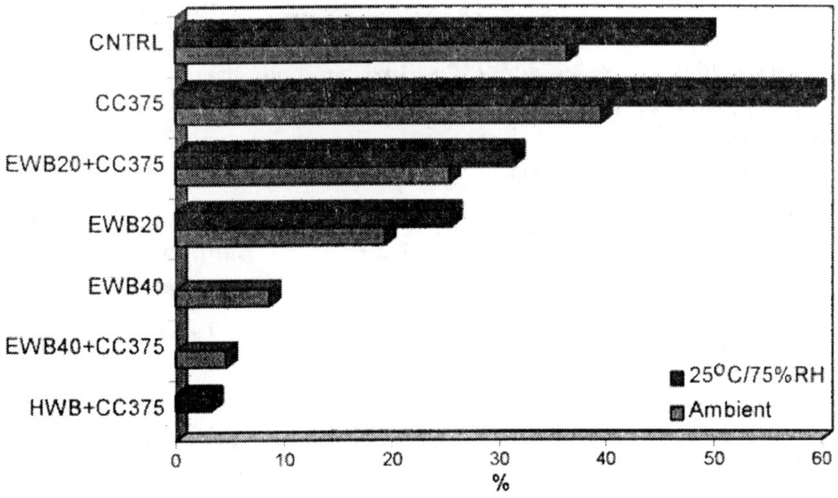

Figure 3.—Percentage weight loss from billets after 8 weeks of storage.

one and 39 percent for trial two, after 8 weeks at 25°C. Such variability was much less pronounced for the control billets. There was no immediate explanation for the variability observed.

Statistical Analysis

Table 3 gives the significant differences (5% level of probability) of sapstain penetration for all the formulations tested for each holding regime.

The salient statistical differences are as follows:

1. After 16 weeks of storage under ambient conditions, EWB40 gave significantly better protection than the commercial standard CC375.
2. After 16 weeks at ambient temperatures, EWB40+CC375 performed significantly better than EWB40, and both performed better than all the other treatments.
3. After 8 weeks of storage at 25°C, HWB+CC375 gave significantly better protection than all the other treatments.

Weight Loss

Untreated control billets had a weight loss of 40.75 percent after 8 weeks storage at 25°C and 75 percent relative humidity (Fig. 3). Billets treated with HWB plus CC375 had a weight loss of 3.23 percent. There is a strong correlation between weight loss and degree of internal sapstain. Those water barrier treatments that were highly effective at reducing the extent of sapstain, including HWB plus CC375, EWB40 plus CC375, and EWB40, also had the lowest weight loss values of 3.23, 4.6, and 8.58 percent, respectively. The maximum weight loss that can occur before significant protection is no longer afforded is not known but these results suggest that it lies between 8.58 percent and 19.26 percent.

Discussion

The results of this experiment suggest that the initial moisture content of radiata pine sapwood is sufficient to prevent sapstain. The possibility that the initial susceptibility to sapstain of different tree species is in part the result of different initial moisture contents is a possibility.

Reference to moisture contents that are expressed as a percentage of the dry weight of the wood are not particularly useful when referring to cardinal moisture contents for fungal growth. This is because it is the percentage of the void volume of the wood occupied by water that is critical and without reference to the wood's density it is not possible to calculate this from percentages of dry weight. Radiata pine sapwood is between 95 percent and 100 percent saturated immediately after felling (Booker, pers. comm.). The results of this study suggest that radiata pine needs to lose 5 percent of it's weight (around 10% of it's wood moisture content) before it starts to become susceptible to sapstain attack.

It is the rate at which water lost from a log during drying (ultimately by evaporation) is replaced by air that is critical in determining the onset of internal fungal colonization. Oxygen has a very limited diffusion potential across water. Since the water in wood cells is largely static in a log, when moisture loss is prevented, the oxygen required for fungal metabolism will only be available in significant concentrations where air is in close contact with the water. Reference in the literature to cardinal oxygen concentrations required for fungal growth in wood are sparse. It seems likely that if a freshly felled (within hours) log is encased in a completely effective water barrier, oxygen concentration will remain at a level that is too low to support fungal growth.

The results showed that for a radiata pine log the rate of radial penetration of sapstain fungal hyphae is greatly reduced when a water barrier is used to prevent moisture loss of approximately 10 percent (approx. 5% of it's weight). However the 10 percent moisture loss is a mean value for the whole log, and the variation in moisture content across the radius of the log is not known. For untreated logs, after a few days of drying a moisture content gradient is well established in stored logs, with the outer sapwood having a lower moisture content than the inner sapwood (Internal FRI report). Pockets of "wet" wood are also common in logs that have been stored for several weeks. It would be expected that a log which has an average moisture loss of 10 percent will have zones of sapwood near the surface of a much lower moisture content than deeper sapwood. However, this may not be the case in logs that have been coated with a water vapor barrier. What could happen is that because moisture is lost at a greatly reduced rate, the movement of water toward the surface occurs at a rate equivalent to the rate of evaporation from the surface. A large moisture gradient would therefore not occur.

The maintenance of a high moisture content in logs may not be the only mechanism of action of a water barrier. In support of this was the observation that cambium coated with water barrier did not develop mould growth as it usually does in the first month of storage. The possible mechanisms whereby a water barrier could have inhibited the mould growth include:

1. natural anti-microbial properties,
2. presence of surfactant adjuvants,
3. acting as an oxygen barrier,
4. acting as a physical barrier.

The significance of the different degrees of protection offered by EWB20, EWB40, and HWB requires consideration. EWB20 and EWB40 are essentially similar formulations, both being water barrier in water emulsions. The higher water barrier content of EWB40 makes the emulsion thicker, although thickness can also be adjusted in other ways. It seems likely that EWB40 performed better than EWB20 because the latter was much thinner (lower viscosity), and very little emulsion adhered to the wood. The thickness of water barrier after the EWB20 emulsion had broken on the logs surface is not known accurately but was estimated to be less than 0.5 mm. The thickness of water barrier on the billets after the EWB40 emulsion had broken was 1 to 3 mm.

A further study (Kreber, internal FRI report) was carried out to investigate the relationship between the rate of water loss and the thickness of the water barrier layer. There was no significant difference between water barrier layers of 0.5 and 3 mm thickness.

Based on a knowledge of molecular structure of water barrier membranes it is known that effective retention of moisture can be achieved with very thin membranes (less than 1 micron). However, the surface of wet wood stored under conditions that promote drying is subjected to forces that are likely to result in rupture of very thin membranes.

1. Drying stresses that result in movement of surface coatings,
2. Physical damage from handling,
3. Temperature induced movement of the water barrier,
4. Absorption into the wood and subsequent dissipation of the barrier.

Furthermore the surface of wood, especially of debarked logs, is very uneven and contains many pits and cavities, the size and shape of which can change with the passage of time. This makes it a difficult substrate to coat with very thin layers of material. It therefore seems likely that very thin layers of a water barrier will not be effective, as was indicated in this study by the difference of performance between EWB20 and EWB40. The surface pits and cavities of wood probably need to be thoroughly clogged with water barrier material before it becomes resilient enough to prevent moisture loss for periods of several months.

The superior performance of HWB compared to EWB40 was not due to the thickness of the barrier, and it is possible that the absence of an emulsion phase resulted in deposition of a water barrier layer that was more hydrophobic.

Detailed isolation studies were not carried out to determine which sapstain fungi were present in logs; however, based on the color and distribution pattern of stain and on subsequent production of perithecia, it appeared that an *Ophiostoma* sp. was predominant in logs stored under both regimes (warm and ambient). *O. piceae* often does not produce bluestain immediately upon colonization of radiata pine logs suggesting that melanization of the hyphae is delayed. Small billet (20 to 25 mm diameter) trials at FRI have shown that under severe drying conditions melanization occurs after 8 to 10 days from the time of inoculation where artificial inoculation occurred 8 hours after felling of the tree (Internal FRI report). Under conditions of slower drying as is usual for larger logs held under ambient conditions, recent observations suggest that total radial penetration of the sapwood can occur without melanization of hyphae occurring. While it has not been proven, there is a strong suggestion that wood moisture content, and probably, concurrently, low oxygen concentration effects the ability of *O. piceae* to undergo melanization in radiata pine.

O. piceae that has colonized sapwood but has not caused bluestain can cause a light brown discoloration. For this reason it is believed that colonization of wood by

O. piceae was not overlooked during the assessments. Furthermore *O. piceae* will usually "bloom" from cross-cut wood faces within 1 week of sawing; and consequently if colonized areas had been overlooked to a significant degree, this would have been observed.

Conclusion

Water barriers can reduce the rate of onset of sapstain in debarked radiata pine logs. The major mechanism of action is believed to be maintenance of an inhibitory oxygen concentration as a result of preventing moisture loss.

Influence of Bark Damage on Bluestain Development in Pine Logs

Dr. A. Uzunovic, Dr. J. F. Webber, and Dr. D. J. Dickinson

Abstract

The use of mechanized harvesters during felling can lead to excessive log debarking, bark loosening, and wood splintering. On average, about a third of the bark was removed from the more severely damaged pine logs used in an experiment which ran from June until October 1993 in Britain. More bark was lost from logs harvested early in the summer (June) than later in the season (August). The main objective of the experiment was to determine whether certain types of damage lead to more bluestain damage than other types. In general, mechanically harvested logs were found to be much more susceptible to attack by bluestain fungi than those processed manually (with a chainsaw). However, certain types of damage were sometimes associated with extensive bluestain development. Bark beetles which act as vectors of some bluestain fungi were excluded from experimental logs, but other insect genera were found to act as casual vectors of staining fungi.

Harvester design and improved harvester operator skills cannot significantly reduce the potential amount of bluestain degrade as significant stain reduction only comes with very low amounts of bark damage (0 to 10% of circumference) which is practically unachievable in practice with mechanized methods. Thus rapid delivery of logs for further processing remains the safest way of minimizing fungal attack.

Dr. A. Uzunovic, Post-Doc Fellow, University of British Columbia, Forintek Canada Corp., Vancouver, BC, Canada, Dr. J. F. Webber, Scientist, Forest Research, Farnham, Surrey, United Kingdom, and Dr. D. J. Dickinson, Senior Lecturer, Department of Biology, Imperial College of Science, Technology and Medicine, London, United Kingdom.

This paper was also published as an IRG Document in 1997.

Introduction

Depending on the conditions and time of storage, freshly felled logs can develop serious deep penetrating bluestain within a few weeks. Among softwoods, which are more prone to stain than hardwoods, pine is known to be especially susceptible. Bluestain is often introduced directly under bark of stored logs via bark boring beetles (8,10). In addition to bark beetle dissemination, bluestain fungi (commonly belonging to genera *Ophiostoma* and *Ceratocystis*) are also vectored via various other insects (1,2,6), water splash, and rarely by wind. Bluestain fungi cannot penetrate directly through bark as they are saprotrophic or only slightly pathogenic; inner wood colonization requires an opening in the bark. Thus intact bark on logs usually gives good and prolonged protection from stain which is not bark beetle transmitted.

During routine forest practices (felling, thinning, extraction, and transportation of logs) bark sustains various degrees of damage and removal. Some mechanical harvesters debark extensively. With the recent increased use of harvesting machines there is rising concern that bark damage caused by harvesters could lead to unacceptable amounts of bluestain. Previous work in Europe showed that logs harvested mechanically were more prone to stain then logs traditionally felled by chainsaw. For example, trials set up in Britain in 1992 (7) confirmed that bark removal was significantly greater with machine harvesters then with chainsaw harvesting; the severity of stain recorded in logs stored for 6 to 12 weeks was significantly greater in machine harvested logs. There were few attempts to quantify the effect of bark damage on stain occurrence and to better understand the underlying process. Earlier observations (4,5,7,9) have all led to the general requirement that mechanized harvesters should cause minimal bark damage and be capable of accurate delimbing.

The study reported here was based on major field experiment set up in England in summer 1993 and 1994.

It evaluated how and to what extent commercial harvesting produced different amounts of bark/wood damage to pine logs and how the amount of damage influenced the extent of bluestain attack. The effect of season and storage time on the occurrence and development of bluestain was also examined. In these trials it was important to exclude any bark beetles as the bluestain introduced by them would mask the effect of harvest damage on stain occurrence. This was done by timing the experiment so that the experimental logs were stored after the local bark beetle flight ended. To exclude possible variation in stain due to the effect of the different design of harvester feeding rollers, the logs used in the experiments were produced by a harvester fitted with rubber feed rollers and selected on the basis of the damage found on them (minimal and maximal damage).

Materials and Methods

Trials were set up on June 24 and August 3, 1993, in Thetford Forest District in the south-east of Britain. Logs were obtained on June 24 and August 3, 1993, from a stand of 64-year-old Corsican pine. On each occasion an Akerman harvester H7 excavator base was fitted with a Lako 60 harvesting head which processed logs between rubber feed rollers with chains produced experimental logs. Logs were 4.2 to 5.2 m in length with top diameters of 15 to 25 cm. On both dates, from several hundred available logs, 20 were selected which had severe bark damage (harvester treatment: maximal damage) and 20 logs with minimal bark damage (harvester treatment: minimal damage). An additional 20 logs were cut with a chainsaw, trimmed, and extracted carefully to avoid any bark damage; these logs were taken as the control treatment. Logs were carefully transferred to a stacking site situated in narrow north-south orientated ridge shaded on both sides by 45-year-old Corsican pine and remote from any felling areas or any other disturbances. A stack 1.5 m high was constructed on bearers with all the logs being aligned east-west. Logs were allocated at random to form batches comprising 5 logs from each of the 3 harvesting treatments and then covered with a layer of extra Corsican pine logs. To further minimize the possibility of the logs being attacked by the bluestain vector bark beetle, *Tomicus piniperda*, small decoy stacks of freshly cut Scots pine were established nearby.

Sampling of Logs

At 3, 6, 9, and 12 weeks a batch of 15 logs (5 of each treatment) was removed for destructive sampling. Discs 3 cm thick were cut, using a chainsaw, 20 cm from the top and bottom of each log. Three more discs were cut at approximately 1.2-m intervals along the remainder of the log. Discs were stored below 5°C and analysed within 3 days of cutting. A large array of measurements were collected from the disks and were grouped into three categories. The first consisted of log parameters comprising disc diameter, bark thickness, and the number of annual rings in the outermost 2 cm of sapwood. The second category consisted of measurements detailing harvester damage such as the amount of bark removed and loosened around the disc circumference, extent of the circumference with the outer bark removed leaving the living phloem exposed, and the extent of the circumference where the exposed wood had been ripped and splintered. The third category of measurements estimated the extent of bluestain colonization and included estimates of the visible stain area on the disc face, the maximum radial stain penetration in the wood, and the extent of stain visible around the circumference of the disc.

Data Analysis

The data were analyzed using analysis of variance (ANOVA). First, the analysis determined if the experimental logs were uniform in terms of diameter, bark thickness, and mean ring width. The extent of bark damage associated with the three treatments was analyzed similarly as well as the three measurements of stain development. A further stage of the analysis explored the relationship between log parameters and harvester damage, between mean annual ring width and stain development, and between the extent of visible stain and bark damage. Lastly the analysis looked into the relationship between stain development and time of sampling, as well as seasonal effects.

Results

Little or no *Tomicus piniperda* or any other bark beetle breeding was visible in the experimental logs, indicating that any bluestain is unlikely to be bark beetle associated. However, various insects (adults or larvae) were collected from the experimental logs. They were identified as species of Coleoptera (*Anaspis* sp., *Arhopalus rusticus*, *Hylobius abietis*, *Pissodes pini*, *Rhizophagus ferrugineus*, *Tharasimus formicarius*), species from the families Elateridae, Staphylinidae, various Collembola, Isopoda; Diptera (Ceratopogimae, Mycetophilidae) and spiders (Arachnoidea). Bluestain damage was visibly greater in both sets of harvester-processed logs compared with the control logs that had been processed with a chainsaw. Indeed, there was no substantial stain development in the control logs even after 12 weeks of storage, and almost all stain visible in these logs could be traced to wounds made by forwarders, which extracted and transported the logs, or had entered through log ends or branch scars.

Damage Caused by Harvesting Process

The different types of damage caused by the chains, delimbing knives, and measuring wheel of the harvester could be seen on the log surfaces together with occasional cracks and splits. However, attempts to consistently identify and estimate the origin of the damage on sampled discs and relate this damage to the stain had to be abandoned. Damage which could be classified and consistently measured included complete bark removal, loosening of the bark still attached to the log, removal of the outer part of the bark which left living phloem tissue exposed, and bruising and ripping of exposed xylem. The most accurate measurements of bark damage came from the 3- and 6-week sampling. Beyond this the bark suffered additional damage as a result of casual insect and fungal activity. Excessive harvester damage was more common on the top logs (logs from the crown) which had live branches and were often pushed through the debranching mechanism of the harvester more then once to ensure complete branch removal.

Table 1 presents data on log diameter, bark thickness, and annual ring width in the outermost sapwood. ANOVA showed that there were no significant differences in these three parameters between experiments and different sampling times. However, there were significant differences in mean diameters ($p > 0.05$), bark thickness, and mean ring width ($p > 0.001$) between treatments, with harvester processed logs with maximal damage tending to be those of smaller diameter (top logs).

Analysis of the extent of bark damage indicated there were significant differences in the amount of bark removed ($p < 0.01$) and loosened (p) and the extent of upper bark removed ($p < 0.01$) between the June and August experiments (Table 2). Not surprisingly, all types of damage were significantly different ($p < 0.001$) between the three treatments in both experiments with very little damage of any kind on the control logs.

For the August experiment, log diameter, bark thickness, and sapwood ring width were not correlated with the extent of bark damage. In contrast, for the experiment set up in June there was a negative correlation between bark loss and bark thickness for the harvester maximal treatment ($p < 0.001$) despite substantial variation ($r^2 = 0.137$), indicating that logs with thicker bark were less likely to suffer harvester damage.

Table 1.—Mean diameter, bark thickness, and annual ring width in logs selected for experiment.

Date	Treatment	Mean diameter	Mean bark thickness	Mean anual ring width
		cm ± SE	*mm ± SE*	
June 21, 1993	Harvester, max	22.5 ±0.28	9.6 ±0.28	2.5 ±0.08
	Harvester, min	24.0 ±0.30	12.4 ±0.41	1.6 ±0.06
	Motor manual	22.5 ±0.35	9.3 ±0.23	2.7 ±0.08
August 3, 1993	Harvester, max	22.0 ±0.27	8.8 ±0.20	2.6 ±0.08
	Harvester, min	23.6 ±0.34	12.4 ±0.44	1.6 ±0.05
	Motor manual	24.0 ±0.34	10.0 ±0.23	2.5 ±0.06

Table 2.—Mean values for the different types of bark and wood damages (measured at 3 and 6 weeks only).

Date	Treatment	Circumference without bark	Circumference with loose bark	Circumference with ripped wood	Circumference with outer bark removed
			% ±*SE*		
June 21, 1993	Harvester, max	34.8 ±2.62	10.6 ±0.96	4.8 ±1.28	5.9 ±0.96
	Harvester, min	12.5 ±1.78	6.0 ±0.90	0.5 ±0.32	5.9 ±0.84
	Motor manual	0.2 ±0.16	0.5 ±0.28	0.0 ±0.00	0.1 ±0.08
August 3, 1993	Harvester, max	29.0 ±2.24	7.8 ±0.79	6.9 ±1.32	3.4 ±0.49
	Harvester, min	6.7 ±1.01	3.7 ±0.71	1.5 ±0.46	1.3 ±0.29
	Motor manual	2.1 ±0.84	1.0 ±0.38	0.2 ±0.16	0.2 ±0.12

Harvester Damage and Development of Bluestain

The relationship between the amount of stain and wood damage was difficult to assess and interpret. To assist in analysis, logs from both harvester treatments were assigned to 10 different categories of damage. The categories started with 10 percent and increased by 10 up to 90 to 100 percent stain area and stain penetration. The stain attributes were measured for last two sampling dates. The area of stain seemed to be positively correlated with greater damage although this was not consistent. For both experiments the area of stain was greater for moderate damage between 30 to 50 percent (June) and 50 to 60 percent (August) then for higher damage in row. Stain penetration did not have a positive correlation with the amount of damaged bark (Fig. 1). By reducing the amount of damage to 30 to 40 percent (June experiment) there was still almost 4 percent of disk area stained; the only significant reduction of stain occurred when damage was reduced to 0 to 10 percent.

Regarding the mean value for stain over time within different treatments, it appears that various amounts of damage (minimal and maximal damage by harvester) in the June experiment produced similar quantities of stain damage for the storage period of 9 weeks. After that there was significant development of stain in logs with maximal harvester damage. The stained area for the August experiment did not progress significantly over time.

Most attempts to isolate various staining fungi within stain areas yielded *Trichoderma* spp. which seemed to be quickly racing through stained wood or even replacing stain fungi (3). Some successful isolations of bluestain fungi yielded *Sphaeropsis sapinea* (42 isolations); it was the most frequently isolated fungus in the June experiment (28 isolations). *Ceratocystis coerulescens* was isolated most frequently throughout the August experiment (14 isolations out of 18). Species of *Leptographium* (*L. truncatum*, *L. procerum*, *L. wingfieldii* and *L.* sp.), *Ophiostoma minus*, *O. piliferum*, *O. piceae*, and *Graphium* spp. were also isolated. The most dramatic staining was produced by *C. coerulescens* and *L. wingfieldii*.

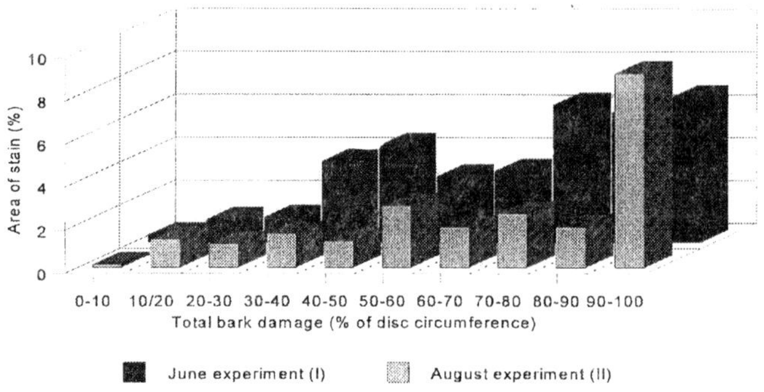

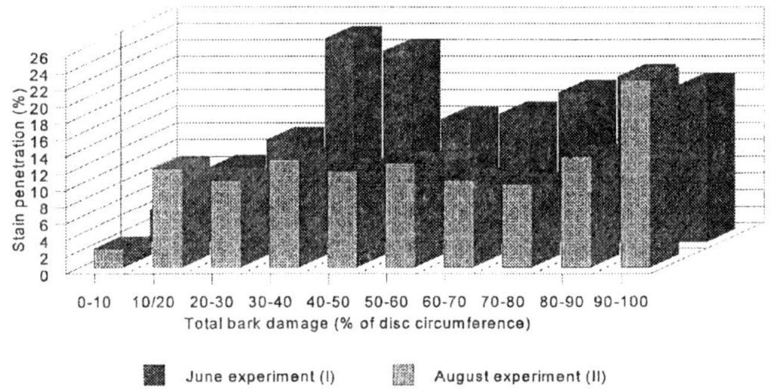

Figure 1.—Amount of stain occuring on logs with varying bark damage.

Discussion

Mechanized harvesting is, at present, the preferred method of felling in many countries and it's future use is very likely to increase. The damage they do on bark and wood can be excessive, and as this work showed, even with less damaging harvesting methods where rubber feed rollers were used, debarking could be as high as 50 percent of the total bark cover. In July the amount of bark damage was greater than latter in the year. This could be explained by the reduced cambial activity later in the summer and, as a result, bark was then more firmly attached to the tree. The amount of produced damage on wood and bark in general can be attributed to various factors, among which the main ones are machine design (knife profiles and knife pressures, design of feeding rollers and their closing pressure), the experience of harvester operator, harvested tree species, position of a log within a tree (top or bottom logs), and time of the year when harvesting was performed. The debarked and damaged areas are usually not continuous but consist of many smaller patches of stripped areas, areas with loosen but still attached bark, crushed or penetrated bark, areas with upper dead part of bark shaved up to living phloem. All this tattered scale of damage present very important and well protected potential infection courts for fungi. These areas are visited by numerous insect fauna which crawl beneath particles of disrupted bark introducing more or less casually associated fungi. The connection between various insects and sapstainers are described in many works. However, this work emphasizes the importance of insects other than bark beetle (especially microarthropods) on bluestain dissemination in logs.

Mechanically harvested logs were clearly far more susceptible to attack by sapstaining fungi than those cut and processed manually by chainsaw. The development

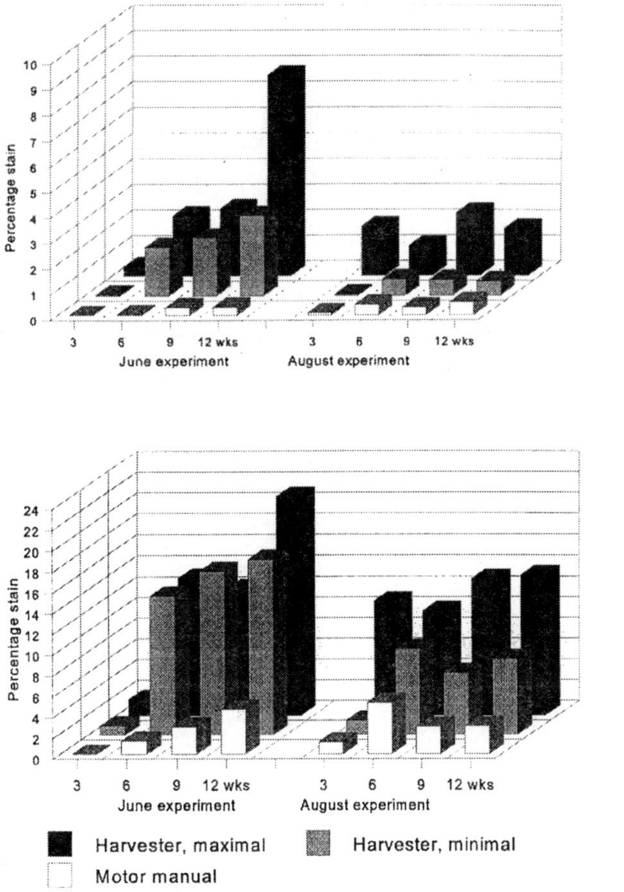

Figure 2.—Development of stain over time: stain expressed as percentage of disc area (top) and stain expressed as percentage radial penetration of disc.

of stain, however, could differ drastically between individual trees within same treatments. The sample of discs cut from mechanically harvested logs which were stored for 9 and 12 weeks (June experiment) revealed about a quarter of them with less then 1 percent stained surface area. The association between various kinds of damage and associated stain at 9 and 12 weeks (when it was maximally developed) was not a simple one (Fig. 2). It was impossible with this experiment to separate various kinds of damages and the resulting stain. Damages from various causes was combined and evaluated as total damage. As a general observation, stain was commonly associated with edges of bark (attached or loosen) and exposed wood. Areas on logs with sizable amounts of stripped bark did not develop a lot of stain. Wood ripping damage did not produce significantly more stain but rather appeared to inhibit it. Where outer bark had been removed and subsequently visited by *Hylobius abietis* which chewed through the areas of exposed phloem and laid its eggs, these areas frequently had stain caused by fungi associated with this beetle (e.g., *Leptographium procerum*).

It is important to note that the average amount of non-bark-beetle-associated stain was, in this work, relatively small (only up to 7.8±1.2% of disk area) even though the experiment was conducted at the optimal time of the year for fungal growth. This is an encouraging result for the foresters and practitioners in Britain. However there is still reason for concern and caution as other timber species and different areas could prove more susceptible. Despite data variability some useful information was acquired and the following conclusion could be made. There was no clear and unreserved evidence that with increasing bark damage greater amounts of bluestain occur. In June that pattern follows partly and the stain damage decreased for disks which had 50 to 70 percent of total damage than disks which had 30 to 50 percent. In August the correlation between increased stain damage and increased wood-bark damage was even less. Stain can develop unexpectedly. Some individual discs in this experiment had stain covering as much as 23 percent of its surface. In south west England there were no grounds for considering that logs from an August felling might be more vulnerable to stain compared with those felled in midsummer but rather opposite, probably due to declining late summer temperatures. As logs age, irrespective of the harvesting treatment, bluestain increases. The rate of degradation appears to decline later in the season probably due to reducing mean temperatures. (The mean monthly temperature for Thetford in June 1993 was 13.9°C and only 9.4°C in October.) There is no supportive evidence that stain would not be problem if harvester damage on logs was small. The results of this work showed that significant reduction of stain may only come with the very low amounts of bark damage (0 to 10%) and in practice this level was not achievable with mechanized harvesting. Thus rapid delivery of logs for further processing remains the safest way of minimizing the opportunity for fungal attack.

Literature Cited

1. Bridges, J.R. and J.C. Moser. 1983. Role of two phoretic mites in transmission of bluestain fungus, *Ceratocystis minor*. Ecol. Entomol. 8:9-12.
2. Dowding, P. 1973. Effects of felling time and insecticide treatment on the interrelationships of fungi and arthropods in pine logs. Oikos 24:422-429, Copenhagen.
3. Gibbs, J.N. 1993. The biology of *Ophiostomatoid* fungi causing sapstain in trees and freshly-cut logs. Chapter 17. In: Ceratocystis and Ophiostoma, taxonomy, ecology and pathogenicity J.M. Wingfield, K.A. Seifert, and J.F. Webber (eds.) APS Press, 293 pp.
4. Helgesson, T. and A. Lycken. 1988. Blånadsskador på virke upparbetat med skördare med slirskyddsförsedda matarhjul av gummi (Blue-stain damage to timber felled by harvester

with non-skid rubber feed rolls). Swedish Institute for Wood Technology Research, Tråteknik-Centrum, Rapport 1, 8801001, 25 pp.

5. Jackobsson, S.G. 1976. Blue-stain damage in mechanically trimmed timber. Meddelande, Svenska-Trarorskningsinstitutet, A. No.390, 36 pp.
6. Leach, J.G., L.W. Orr, and C. Christiansen. 1934. The interrelationship of bark beetles and blue staining fungi in felled Noway pine timber. Jour. Ag. Res. 49(4):315-341.
7. Lee, K. and J.N. Gibbs. 1996. An investigation of the influence of harvesting practice on the development of blue-stain in Corsican pine logs. Forestry 69(2):129-133.
8. Mathiesen-Käärik, A. 1953. Eine übersicht über die Gewöhnlichsten mit Borkenkäfern assoziierten Bläuepilze Schweden und eining für Schweden neue Bläuepilze. Meddn. St. Skogsforsk. Inst. 43:3-74.
9. Sderstrm, O. 1986. Unprotected storage of coniferous saw logs in Sweden. Report No 172. The Swedish Univ. of Agric. Sciences.
10. Wingfield, M.J., P. Capretti, M. Mackenzie. 1988. *Leptographium* spp. as root pathogens of conifers. An international perspective. In: *Leptographium* root diseases of conifers (T.C. Harrington and F.W. Cobb, Jr., eds.), APS Press, St. Paul, Minnesota. pp. 113-128.

Defacement of Freshly Sawn Corsican Pine Lumber by Sapstain and Mould Fungi and the Influence of Arthropods

N. J. Strong, Dr. J. F. Webber, and Dr. R. A. Eaton

Abstract

In a trial to study the arthropods associated with sapstain and mould fungi, freshly sawn Corsican pine (*Pinus nigra* var. *maritima* (Aiton) Melville) boards were block stacked at a local sawmill. Each stack was constructed in a nest arrangement with smaller boards, in experimental tanks, positioned in the center of the stacks. Boards were converted from a stock of sawlogs at monthly intervals, to provide host material of different ages. Filters and screens were used to exclude arthropods and airborne inoculum from selected experimental tanks. The experimental tanks were removed monthly, and the insect community and fungal defacement assessed. The highest numbers of arthropods were recorded during the summer months, maximum numbers being approximately 20 insects and 100 mites per 500 cm^2 board surface. Ninety-four species of arthropods have been identified from the experimental material (sawlogs through to experimental boards) with 90 percent from within the stacked lumber itself. The most abundant were Dipteran larvae, fungus gnats (Mycetophilidae: Diptera), and *Atheta coriaria* Kraatz (Staphylinidae: Coleoptera). The exclusion of the arthropods from experimental tanks lead to a significant rise in the amount of mould, but not sapstain fungi. Studies were carried out at different seasons over 2 years. The significance of arthropods in the fungal defacement of sawn lumber is discussed.

N. J. Strong, School of Biological Sciences, University of Portsmouth, Portsmouth, United Kingdom, Dr. J. F. Webber, Forest Research, Wrecclesham, Farnham, Surrey, United Kingdom, and Dr. R. A. Eaton, School of Biological Sciences, University of Portsmouth, Portsmouth, United Kingdom.
This paper was also published as an IRG Document in 1997.

Introduction

There are three groups of fungi generally involved in the defacement of timber (18). The Ophiostomatoid fungi, which include species of *Ceratocystis*, *Ophiostoma*, and *Ceratocystiopsis*, are penetrating stainers whose hyphae penetrate into the rays and tracheids. The second group incorporate the black yeasts such as *Hormonema dematoides*, *Aureobasidium pullulans*, *Rhinocladiella atrovirens*, and *Phialophora* spp. These staining fungi cause superficial stains on the surface of the timber. Mould fungi such as the dark moulds are included in the third group. The dark moulds include *Alternaria alternata* and *Cladosporium sphaerospermum* which produce abundant conidia and superficially stain the wood. These moulds also comprise *Penicillium* sp. and *Trichoderma* sp. which frequently occur on sapwood and cause green discoloration.

Where fungal decay occurs prior to processing, the wood can be discarded before use, but after this any staining that occurs has many implications for the end use of the timber. Although the damage caused is usually cosmetic and the mechanical properties are minimally affected (2), the stain does cause the timber to be down graded (British Standard 3819: 1964) and this leads to a decrease in value (e.g., 1,7,10,16,17).

It is often assumed that the fungal infestation of freshly cut wood surface is by aerial dissemination of fungal spores or infected sawdust particles. The role of insects as additional vectors in this process, however, has been presented by Powell et al. (16). The general relationship between insects and fungi is well documented (e.g., 13,14,21) especially in terms of dependency on each other for food and dispersal. The insect dissemination of propagules has been reported by Verrall (19) and Dowding (5) though this is often casual dispersal and there are species, especially of the Scolytidae: Coleoptera, that carry spores in specialized mycangia which inoculate the surface of galleries in wood and provide food for the larvae (6). Bridges and Moser (2) described a bark beetle (*Dendroctonus frontalis* Zimm. (Scolytidae: Coleoptera)) with phoretic mites that are hypervectors of the bluestain fungus, *Ceratocystis minor*. Fungivorous insects are represented by a wide range of orders, the most common being Coleoptera, Diptera, Acari, and nematodes, which feed on all parts of the growing fungus from mycelium to spores (6).

There are two main methods currently used in an attempt to control the effect of the sapstain and mould fungi. Kiln-drying of the lumber reduces the moisture content of the wood from 150 to 200 percent to an ideal level of 26 to 28 percent and prevents the growth of most fungi (20). Fungicidal spraying or dipping of timber in applications containing copper-8-quinolinolate does prevent some growth but due to non-specificity and great variation between target species their efficacy is reduced (4,11). Evidence for the natural biological control of fungi by arthropods was reported by Powell et al. (16) who stated the incidence of mould on sawn lumber to be negatively correlated to the numbers of mites and insect larvae and Dowding (6) who established that the mycelial mats of *Ceratocystis coerulescens* and *C. pilifera* on the crosscut ends of logs were shredded by mites after exposure for a week.

The ability of the fungi to invade host cells is determined by the condition of the timber, both in terms of health and age. Gibbs (8) identified a range of fungi from the truly pathogenic species occurring in living trees to those colonizing weakened or stressed trees, to the saprobic fungi on dead timber. *Ceratocystis coerulescens* has been isolated from a stand of living Norway spruce (*Picea abies*) in Sweden, but in Europe the same species, although recorded in recently cut logs of Scots pine (*Pinus sylvestris*), has not been found in healthy trees (8). Similarly Dowding (6) stated that the mould fungi *Trichoderma* spp. are unable to invade live wood. Powell et al. (16) showed that sapstain defacement of freshly sawn, untreated Corsican pine boards reached 90 percent coverage after 8 weeks of exposure, and Keirle (12) showed that species of staining fungi could be isolated from new boards after 1, 2, or 3 months pre-storage as logs in a forest in Australia.

The investigation detailed in this paper was carried out as a set of three trials over a period of 2 years and encompassed the different growing seasons. The conversion of sawlogs to boards at staggered intervals over the trial periods gave rise to different aged timber being exposed to the fungi and arthropods over these seasons. This has enabled the effect of timber condition and season on the infestation of sawn lumber to be described. The experimental design involves the use of physical barriers to gather evidence for the role of arthropods in the dispersal and possible biological control of fungal staining.

Materials and Methods

Trial Location and Log Conversion

Sawlogs used for the experiment were Corsican pine (*Pinus* nigra var. *maritima* (Aiton) Melville) approximately 4.0 meters in length with a minimum top diameter over bark of 18 centimeters. The logs were stored in a separate pile at a sawmill in Hampshire, UK, but close to other logs and converted lumber of different ages and species (both hard and softwood). The sawlogs were converted to boards at the following time intervals; 0 (i.e., freshly felled), 4, 8, and 16 weeks. The boards sawn from the sawlogs measured 1.0 m by 90 mm by 18 mm and 0.45 m by 90 mm by 18 mm and were used as the outer casing timber for the construction of nested stacks. Boards measuring 0.25 m by 90 mm by 18 mm were placed in experimental tanks and positioned in the center of each stack. Any boards with visible fungal infection, a high proportion of heartwood, or a significant wany edge were discarded prior to stack construction. The trial was repeated three times with trees felled in June 1995, December 1995, and September 1996.

Stack Construction and Experimental Tanks

The stacks measured 1.0 m long by 0.81 m wide and approximately 1.0 m in height. A nest arrangement similar to that used by Powell, Eaton, and Webber (15) and Powell et al. (16) was employed and two experimental tanks were placed within each stack (Fig. 1). Heavy duty

Figure 1.—Experimental stack under construction with tanks in nest area.

plastic sheeting was used as a rain roof to prevent excess wetting and each stack was built on a clean wooden pallet to lift the experimental boards off the ground and to enable the finished stack to be moved easily by fork-lift truck.

Plastic tanks (20 by 34 by 22 cm) were tightly packed with 0.25 m long experimental boards. Each tank contained 17 boards and the tanks were left open or were closed. The closed tanks were sealed with a piece of Perspex which was held in place using waterproof silicone sealant. A 100 cm^2 area at the center of the Perspex was cut out and covered with a filter membrane ("Fluorotrans" transfer membrane, Pall BioSupport, Hampshire, UK). The pore size of the membrane was 0.2 µm which allowed gaseous exchange but prevented the entry of any arthropods or airborne fungal spores into the tank. The membrane filters were protected by a nylon mesh (approximately 2 mm^2 gauge) covering the whole face of the tanks, and the tanks were placed in the nest on their sides with the open, or filter covered, face towards the side of the stack.

The boards inside the open tanks were used as 'bait' for the insects so that the faunal community associated with stacked lumber could be determined. The open tanks were used in conjunction with the closed tanks to identify any relationships between the insects and the fungal species that colonized both tanks. The closed tanks were used to identify any contamination that occurred between milling and the construction of the stacks.

Trial Period

Fifteen sawlogs were converted at each time period providing enough boards to enable two stacks to be built. The conversion of logs at intervals over the course of the trials meant that at any one time the number of experimental stacks at the sawmill varied between 2 and 8. These stacks, at a spacing of approximately 1.0 m, were situated alongside commercial lumber used for pallet-making which was often quite badly defaced. Experimental tanks were removed from the stacks after 4 weeks of exposure. The open tanks were sealed immediately on removal from the nest structure to prevent escape of any arthropods. Both types of tank were kept at 4° to 8°C pending assessment for the presence of arthropods, fungal defacement, and moisture content determination of the experimental boards.

Converting the logs at weeks 0, 4, 8, and 16 and then exposing the lumber for 4 weeks prior to inspection provided experimental material aged between 4 and 20 weeks after felling. This allowed comparisons between the sealed and open tanks, between different aged timber exposed in stacks for the same length of time, and between different times of the year.

Defacement, Moisture Content, and Arthropod Assessment

Overall defacement of experimental boards in each tank was recorded photographically prior to removal from the tanks and detailed visual assessments. Defacement was determined as the percentage surface area of each experimental board covered by sapstain, mould, and basidiomycete fungi.

Cores of wood, 30 mm in length and 11 mm in diameter, were taken from 10 of the boards randomly selected from within the experimental tanks in order to calculate residual moisture content.

The upper surface of each of the sample boards from the experimental tanks was examined for arthropods using a 25× binocular microscope. Adult insects and larvae were removed and preserved in 70 percent IMS (industrial methylated spirits). The mite population was determined by sampling 15 cm^2 of the board surface and counting the number of mites found. Identification of the insects to at least family level was carried out following preservation in the IMS.

Weather Conditions, June 1995 through February 1997

The months of 1995 in which the first trial was carried out were generally warmer than the previous 16 years average; the months of June through August were considerably drier and conversely September was much wetter than the average. In contrast, 1996 was cooler and drier than the average of the previous 16 years, though November was the wettest in 15 years. The temperature and rainfall data are shown in Figure 2.

Results

The results of moisture content, fungal defacement, and arthropod abundance are derived from the experimental ("bait") boards contained in tanks; information on the casing boards is not presented below and will be presented elsewhere.

Moisture Content Determination

The moisture content values of the experimental boards for the three trials are given in Figure 3. There was no significant difference between the moisture content determinations of the experimental boards from the tanks with open faces and from those with the covering filter (two-way ANOVA; $F_{1,36} = 4.11, P > 0.01$). After 4 weeks of exposure the moisture content of the boards inside the tanks, in most cases, was not significantly different from the boards immediately after conversion (two-way ANOVA; $F_{1,36} = 4.11, P > 0.01$). However, the moisture content determinations for November 1996 increased

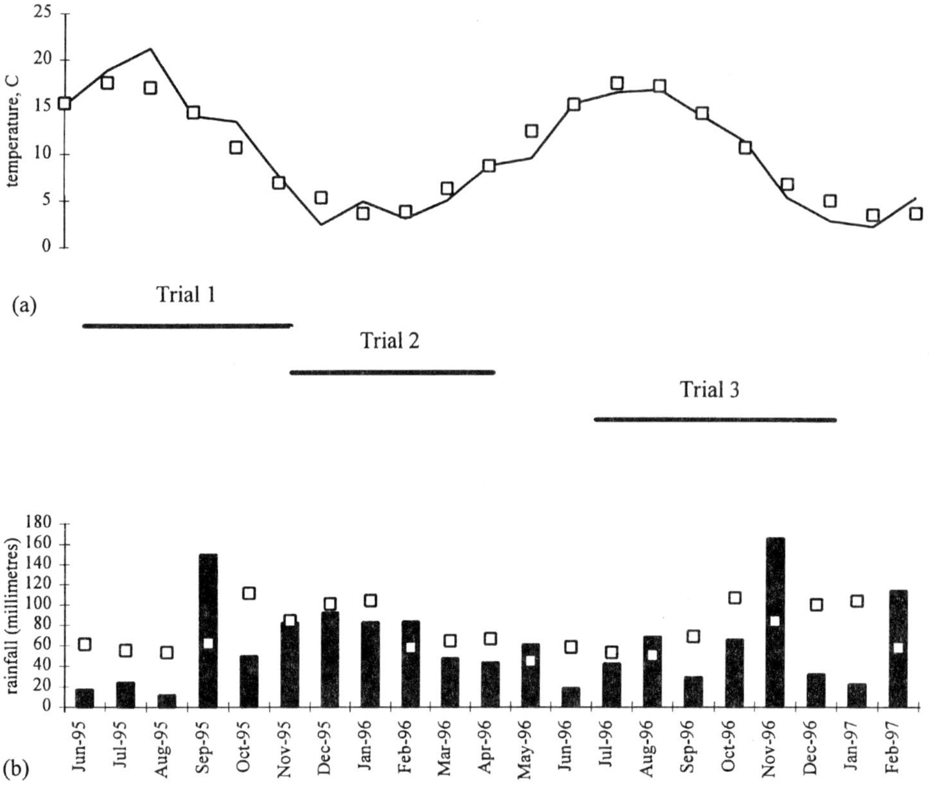

Figure 2.—Mean monthly temperatures and rainfall during the three trial periods: (a) temperature (°C): 1995-1997 data, —; previous 16 years average, □. (b) rainfall (mm): 1995-1997 data, ■; previous 16 years average, □. (Courtesy of K.F. Hoad, Widley, Hampshire, UK.)

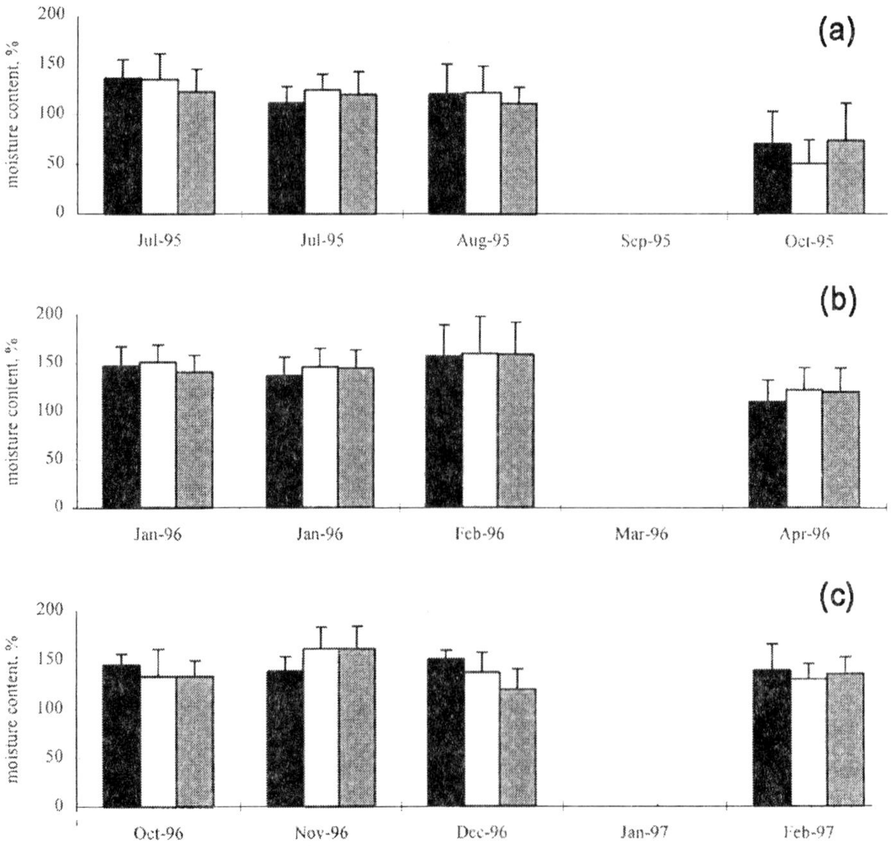

Figure 3.—Mean moisture contents, calculated as percentage dry weight, (± standard deviation) for the experimental bait boards at time of conversion and after 4 weeks of exposure. (a) Trial 1. (b) Trial 2. (c) Trial 3. Key: Boards at week 0, ■; boards at week 4: filter-covered tanks, □; open-faced tanks, ▩.

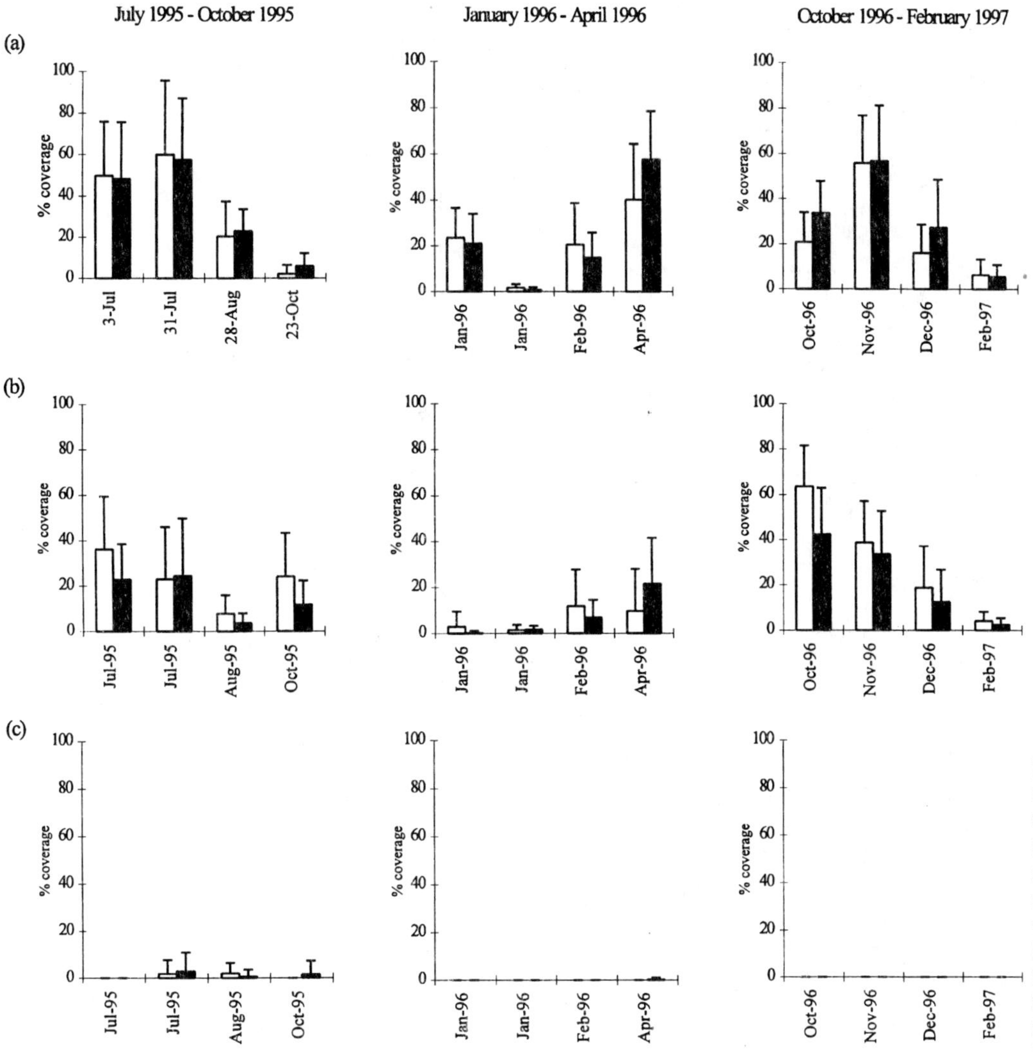

Figure 4.—Defacement of experimental boards from within sample stacks given as a percentage of board surface area covered (± standard deviation) after 4 weeks of exposure. (a) Sapstain fungi. (b) Mould fungi. (c) Basidiomycete fungi. Key: boards from filter-covered tanks, □; boards from open-faced tanks, ■.

from 138 percent to 160 percent, correlating with extremely heavy rainfall in this month.

Fungal Defacement of Experimental Boards

The maximum levels of defacement of the experimental boards by sapstain and mould fungi were 60 to 65 percent of the board surface after 4 weeks of exposure in tanks. Colonization by the basidiomycete fungi remained at very low levels throughout the trials with a maximum coverage of 5 percent. Figure 4 shows the levels of sapstain, mould, and basidiomycete infestation for the three trial periods. The intensity of infestation follows a seasonal trend, higher over the summer months and lower through the winter.

The extent of colonization of experimental boards by sapstain fungi in open and closed tanks was significantly different only on three occasions: October 1995, April 1996, and October 1996. On these occasions the levels of sapstain defacement over the boards were greater in the open tanks (two-way ANOVA; $F_{1,64} = 3.99$, $P < 0.01$) than the filter covered tanks. Despite the fact that the exposure period was the same for all the sawn material the levels of sapstain colonizing the experimental boards cut from freshly felled sawlogs was usually less than that seen on the boards from 4-week-old sawlogs in Trials 1 and 3. This suggested that age and condition of the host material influences levels of fungal colonization on the wood surface.

Intensity of infection by mould fungi differed in the two types of experimental tank. In general, colonization by moulds was greater in the tanks with the filter cover. These differences were statistically significant in July 1995, October 1995, and October 1996 (two-way ANOVA; $F_{1,64} = 3.99$, $P\ 0.01$). There was no significant difference between the intensity of mould growth on boards in open and closed tanks for the whole of Trial 2,

	July - October				January - April				October - February			
	Age of timber at sampling				Age of timber at sampling				Age of timber at sampling			
	4	*8*	*12*	*20*	*4*	*8*	*12*	*20*	*4*	*8*	*12*	*20*
LARVAE												
Diptera												
Mycetophilidae	**12**	**39**	**50**	**60**	0	0	0	0	**70**	**63**	0	0
Brachycera	**1**	**1**	0	0	0	0	0	0	**3**	**2**	0	0
Ceratopogonidae	**1**	0	**18**	0	0	0	0	0	**83**	0	0	0
Ceratopogonidae	0	0	**9**	0	0	0	0	0	0	0	0	0
Silvicolidae	**9**	**1**	0	0	0	0	0	0	0	0	0	0
Coleoptera												
Aleocharinae	**1**	**2**	**8**	**4**	0	0	0	0	0	0	0	0
Aleocharinae	0	0	**12**	**1**	0	0	0	0	0	0	0	0
Chrysomelidae	**4**	**1**	0	0	0	0	0	0	0	0	0	0
Chrysomelidae	0	**1**	0	**1**	0	0	0	0	0	0	0	0
Staphylinidae	0	**1**	0	0	0	0	0	0	0	0	0	0
unknown	**1**	0	**3**	0	0	0	0	0	0	0	0	0
PUPAE												
Diptera												
Mycetophilidae	0	**12**	**5**	**2**	0	0	0	0	**13**	0	0	0
unknown	**1**	0	**18**	0	0	0	0	0	0	0	0	0
ADULTS												
Diptera												
Mycetophilidae	**3**	**9**	**14**	**1**	0	0	0	0	0	**10**	0	0
Acartopthalamus sp.	0	**1**	0	0	0	0	0	0	0	0	0	0
Cecidomyiidae	0	**1**	**8**	0	0	0	0	0	0	0	0	0
Cecidomyiidae	0	0	**6**	0	0	0	0	0	0	0	0	0
Drosophilidae	**1**	0	0	0	0	0	0	0	0	0	0	0
Mycetophilidae	0	**1**	0	0	0	0	0	0	0	0	0	0
unknown	**1**	**1**	0	0	0	0	0	0	0	0	0	0
Coleoptera												
Atheta coriaria	**5**	**7**	**8**	**4**	0	0	0	0	0	0	0	0
Aridius species	**1**	**1**	0	0	0	0	0	0	0	0	0	0
Epuraea pusilla	0	0	0	0	0	0	0	**1**	0	0	0	0
Falagria concinna	0	0	**1**	0	0	0	0	0	0	0	0	0
Lathridius anthracinus	0	0	0	0	0	0	0	0	0	0	**1**	0
Phloeonomus pusillus	0	0	0	**5**	0	0	0	0	0	0	0	0
Hymenoptera												
Braconidae	**1**	0	0	0	0	0	0	0	0	0	0	0
Sphecidae	0	**1**	0	0	0	0	0	0	0	0	0	0
Hemiptera												
unknown	0	**1**	0	0	0	0	0	0	0	0	0	0
unknown	0	**1**	0	0	0	0	0	0	0	0	0	0

Table 1.—Insect numbers collected from the experimental tanks after 4 weeks of exposure. Data from three trials.

but overall levels of colonization were very low during this period. Levels of mould defacement do appear to show a seasonal trend, as with the sapstain, though this was lower in August 1995 and corresponded with the meteorological data when August was particularly warm and dry.

Assessment of Arthropod Populations

A total of 94 species of arthropod (excluding mite species) representing 14 orders were collected from stacks and logs during the course of the three trials. Eighty-three species were found within the stacked lumber, 19 of which were also caught within the "bait" tanks. The majority of arthropods were captured during the summer in Trial 1. Trial 3 had the next highest count and only 1 insect was found throughout the whole of Trial 2. Most numerous were the insect larvae, a mixture of Dipteran, Coleopteran, and Neuropteran species. Of the adult insects captured the most common were a fungus gnat (Mycetophilidae: Diptera) and a rove beetle (*Atheta coriaria* Kraatz (Staphylinidae: Coleoptera)). Relative abundances of the insects captured in the experimental tanks are displayed in Table 1 and a complete species list is give in Table 2. *Ips sexdentatus* Boerner (Scolytidae: Coleoptera) was the most common insect on the sawlogs. In Trial 1, on the 16-week-old logs it was estimated that the density of beetles in galleries under the bark was 1 beetle per 25 cm^2. Mite numbers were calculated per 500 cm^2 of board surface area. No mites found during Trials 2 and 3, and the numbers present on the lumber exposed for 4 weeks during Trial 1 peaked on the 8-week-old timber with approximately 70 mites per 500 cm^2 decreasing to 9 mites on the 20-week-old wood.

Order	Taxon
Diptera	**Mycetophilidae**
	unknown species
	Mycetophilidae
	Cecidomyiidae
	***Acartopthalamus* species**
	Drosophilidae
	Cecidomyiidae
	Anisopodidae
	Cecidomyiidae
	Clusiidae
	Clusiinae
	Dolichopididae
	Dolichopus griseipennis Stannius
	Muscidae
	Mycetophilidae
	Mycetophilinae
	Nephrotoma appendiculata
	Periscelidae
	Platypezidae
	Scatopsinae
	Sciomyzidae
	Sciophila hirta Meigen
	Silvicolidae
Coleoptera	*Aridius nodifer* Westwood
	***Atheta coriaria* Kraatz**
	***Phloeonomus pusillus* Gravenhorst**
	Atheta species
	***Epuraea pusilla* Illiger**
	Ips sexdentatus Boerner [†]
	Pissodes castaneus Degeer [†]
	Tomicus piniperda L. [†]
	***Aridius* species**
	Cypha longicornis Paykull
	Scolytidae
	Aridius bifasciatus Reitter
	Athous bicolor Goeze
	Atomaria atricapilla Stephens
	Coccinella 7-punctata L. [†]
	Cryptophagus dentatus Herbst
	Cylindrinotatus laevioctostriatus Goeze [†]
	Dromius quadrinotatus Zenker in Panzer [†]
	***Falagria concinna* Erichson**
	Hylastes opacus Erichson [†]
	***Lathridius* species**
	Lathridius anthracinus Mannerheim
	Leperisinus varius Fabricius
	Phyllobius pomaceus Gyll. (=*urticae* (Degeer))
	Phyllotreta nigripes Fabricius [#]
	Proteinus brachypterus Fabricius
	Scolytidae [†]
	Sepedophilus littoreus L.
	Silvanus bidentatus Fabricius
	Sitona lineatus L.
	Tytthaspis 16-punctata L.
Hymenoptera	**Braconidae**
	Diapriidae
	Braconidae *
	Figitidae *
	Figitidae
	Formicinae
	Halictinae
	Ichneumonidae *
	Ichneumonidae
	Sphecidae
	Sphecidae/Miscophidae
	unknown species
	Vespula vulgaris
Collembola	*Tomocerus vulgaris* (Tullberg)
	Lepidocyrtus cyaneus
	Entomobrya albocincta
	Entomobrya nivalis (L.)
	Lepidocyrtus curvicollis
	Orchesella atticola (Uzel)
	Orchesella species
	Tomocerus longicornis (Muller)
	Tomocerus minor (Lubbock)
Hemiptera	Acucephalinae
	Agalliidae
	Anthocoridae
	unknown species
	unknown species
Dermaptera	*Forficula auricularia*
Lepidoptera	unknown species
Orthoptera	Acrididae
Thysanoptera	*Phlaeothrips annulipes*
Neuroptera	unknown larva
Araneae	*Tegenaria gigantea*
	Araneae
	Araneae
	Araneae
	Zodariidae
Diplopoda	unknown species
Chilopoda	*Lithobius variegatus*
Isopoda	*Porcellio scaber*

Table 2.—Species list of arthropods (excluding mites) captured over 3 trials. Insects were captured within the nested stacks (normal font), within the experimental tanks (**bold** font), on the sawlogs (†) or in the sawmill yard (*). # denotes species found on sawlogs and within nested stack.

Insects present within the stacks were not necessarily associated with either fungi or timber. For example, *Phyllotreta nigripes* Fabricius (Chrysomelidae: Coleoptera) feeds on members of the Brassicacae and species such as *Cypha longicornis* Paykull (Staphylinidae: Coleoptera) is a predator of mites. *Falagria concinna* Erichson (Staphylinidae: Coleoptera), an Asian species, possibly originated from the exotic timber dealt with by the sawmill; this record maybe only the second in the British Isles (P. M. Hammond, pers. comm.).

Discussion

The average moisture content of the sample boards within the experimental tanks did not fall below 50 percent in either the open or closed tanks. Fungi are able to colonize timber at moisture contents of between 30 percent and 150 percent, with an optimum of 60 to 80 percent (e.g., 10,18). Above or below these levels the wood is either too wet or too dry for optimum growth conditions. The lumber under investigation over the 4 weeks of exposure was constantly susceptible to growth by the defacing fungi. The lack of significant differences between the moisture content determinations for the two types of tank and between the freshly converted boards and the lumber exposed for 4 weeks suggests that any differences in levels of defacement were due to other variables.

Defacement on boards in the filter-covered tanks indicates considerable contamination in the time interval between conversion of the sawlogs and stacking of the lumber. In each of these trials this time interval was invariably less than 2 hours. This contamination may originate from the saw blade and wood particles containing viable fungal propagules. It also cannot be ruled out that some of the boards were already colonized by sapstainers which were not yet pigmented. Although the arthropods are known to have a role in the dissemination and control of the defacing fungi of sawn timber, when they are excluded high levels of defacement still result because of inoculum dispersal.

Despite this, there were significant differences between the amount of sapstain defacement in the two tanks, with more sapstain occurring when insects had access to the tanks, and when insects were most abundant. Mould fungi with drier spores are easily dispersed in air currents (15), although they can also be transmitted by casual insect dispersal. Levels of mould in the experimental tanks with filter covers preventing insect entry were significantly greater than those in the open tanks. This was recorded only in the warmer months when there were insects captured in the bait tanks. The majority of the arthropods captured in the bait tanks are described in the literature as fungivores (e.g., 9), and their abundance might explain the differences observed in the trials. This link between the presence (or absence) of the arthropods and the levels of defacement over the surface of the boards agrees with work of Powell et al. (15,16), i.e., the arthropods act as vectors bringing inocula to the boards but can then also graze the moulds to the extent that levels are increased or reduced.

There is some evidence to suggest that the age of the timber has an influence on the amount of deterioration that occurs. The older timber has higher levels of disfigurement by sapstain and mould fungi than timber from logs converted weeks earlier. However, the increasing amounts of defacement visible on the lumber, especially caused by sapstain species on 20-week-old timber, is often masked as the amount decreases due to seasonal or weather changes. Figure 4 clearly shows the lower levels of defacement occurring at the end of the year, followed by a subsequent rise with the onset of spring weather. Extreme conditions can also be seen to have an effect, as shown by the decrease in both sapstain and mould fungi in August 1995, corresponding to unusually warm and dry weather.

The lower levels of defacement on freshly felled timber boards and boards processed in the winter months point to the value of either immediate conversion from sawlogs to end material, or log storage until winter then milling. Storage and then sawing in the warmer months compounds the problem of defacement as shown by the increased amounts of sapstain and mould on the 20-week-old timber in April 1996.

Further investigations into the age and condition of the wood, as indicators of host resistance, in conjunction with the effects of weather and season, need to be carried out to ascertain any methods of reducing the levels of defacement.

Acknowledgments

The authors wish to acknowledge the Forestry Commission, UK and the University of Portsmouth for their financial support during these trials. Thanks to the staff at East Brothers (Timber) Ltd, West Dean, nr. Salisbury, Hampshire, UK for their help, facilities and patience in the setting up of the field trials. We also wish to thank Peter Hammond (Natural History Museum), Tim Winter (Forestry Commission), Brian Cave (Longhope, Gloucestershire) for invaluable help in the identification of the insects.

Literature Cited

1. Anonymous. 1955. Sap-stain in timber: its cause recognition and prevention. Dept. Sci. Ind. Res. Leaflet no. 2. For. Prod. Res. Lab., Princes Riseborough, England.
2. Bridges, J.R. and J.C. Moser. 1983. Role of two phoretic mites in transmission of bluestain fungus, *Ceratocystis minor*. Ecological Entomology 8:9-12.
3. Cartwright, K.St.G. and W.P.K. Findlay. 1958. Decay of timber and its prevention. HMSO, London. pp. 298-320.
4. Dickinson, D. J. 1988. Recent problems and developments in the control of post-harvest deterioration of softwood timber by chemical treatments. International Biodeterioration 24(1):321-326.
5. Dowding, P. 1969. The dispersal and survival of spores of fungi causing bluestain in pine. Transcripts of the British Mycological Society 52(1):125-137.
6. Dowding, P. 1984. The evolution of insect-fungus relationships in the primary invasion of forest timber. In: Invertebrate-Microbial Interactions. J.M. Anderson, A.D.M. Rayner, and D.W.H. Walton (eds.). Cambridge University Press, Cambridge. pp. 134-153.
7. Findlay, W.P.K. 1959. Sapstain of timber: part II. Forestry Abstracts 20(2):167-174.
8. Gibbs, J.N. 1993. The biology of ophiostomatoid fungi causing sapstain in trees and freshly cut logs. In: Ceratocystis and Ophiostoma: Taxonomy, Ecology and Pathogenicity. M.J. Wingfield, K.A. Seifert, and J.F. Webber (eds.). The American Phytopathological Press, St. Paul, Minnesota. pp. 153-160.
9. Hammond, P.M. and J. F. Lawrence. 1989. Mycophagy in Insects: a Summary. In: Insect-fungus interactions. N. Wilding, N.M. Collins, P.M. Hammond, and J.F. Webber (eds.). Academic Press, London. pp. 275-324.
10. Holtam, B.W. 1966. Blue stain—A note on its effect on the wood of home grown conifers and suggested methods of control. Forestry Commission Leaflet 53. p 4.

11. Kay, S.J. 1995. The biological control of sapstain of Pinus radiata with microorganisms. Unpublished PhD thesis, University of Auckland. p 185.
12. Keirle, R.M. 1978. Effect of storage in different seasons on sapstain and decay of Pinus radiata D. Don in N.S.W. Australian Forestry 41(1):29-36.
13. Leach, J.G., L.W. Orr, and C. Christiensen.1934. The interrelationship of bark beetles and blue staining fungi in felled Norway pine timber. J. Agric. Res. 49(4):315-341.
14. Malloch, D. and M. Blackwell. 1993. Dispersal biology of the ophiostomatoid fungi. In: Ceratocystis and Ophiostoma: Taxonomy, Ecology and Pathogenicity. M.J. Wingfield, K.A. Seifert, and J.F. Webber (eds.). The American Phytopathological Press, St. Paul, Minnesota. pp. 195-206.
15. Powell, M.A., R.A. Eaton, and J. F. Webber. 1994. Insect transmission of fungal sapstain to freshly sawn unseasoned softwood lumber. The International Research Group on Wood Preservation Document No. IRG/WP/94-20025.
16. Powell, M.A., R.A. Eaton, and J. F. Webber. 1995. The influence of micro-arthropods on the defacement of sawn lumber by sapstain and mould fungi. Canadian Journal of Forest Research 25:1,148-1156.
17. Savory, J.G., R.G. Pawsey, and J.S. Lawrence. 1965. Prevention of blue-stain in unpeeled Scots pine logs. Forestry 38:59-81.
18. Seifert, K.A. 1993. Sapstain of commercial lumber by species of Ophiostoma and Ceratocystis. In: Ceratocystis and Ophiostoma: Taxonomy, Ecology and Pathogenicity. M.J. Wingfield, K.A. Seifert, and J.F. Webber (eds.). The American Phytopathological Press, St. Paul, Minnesota. pp. 141-151.
19. Verrall, A.F. 1941. Dissemination of fungi that stain logs and lumber. Journal of Agricultural Research 63:549-558.
20. Wakeling, R.N. 1988. The control of mould and sapstain on freshly felled timber. Unpublished MSc thesis, Portsmouth Polytechnic. p 120.
21. Wilding, N., N.M. Collins, P.M. Hammond, and J.F. Webber (eds.). 1989. Insect-fungus interactions. Academic Press, London. p 344.

Biological Control: Panacea or Boondoggle

J. J. Morrell and B. E. Dawson-Andoh

Abstract

Biological control (biocontrol) has received tremendous interest from a variety of industries seeking to reduce their dependence on synthetic pesticides. Biocontrol, however, is not without its difficulties. An excellent example is in the area of preventing stain and mould on freshly sawn lumber. While excellent results have been produced in laboratory trials, these results have not translated into field success. This report summarizes the problems associated with biocontrol of stain fungi, identifies strategies for overcoming these problems, and outlines research needs for improving the prospects for commercial biocontrol.

Introduction

Freshly sawn lumber remains susceptible to attack by a variety of fungi until its moisture content declines below 20 percent (wt/wt) (45). While many fungi can impact wood properties, the fungi that discolor the surface or the interior of the finished product are of critical concern since they can significantly impact the value of the final finished product. For example, values for clear, stain-free ponderosa pine can be five times that of stained lumber.

For many years, mills controlled or prevented stain by topical application of fungicides shortly after sawing (38,39). The primary chemical used for this purpose was sodium pentachlorophenate (penta) a broad spectrum biocide that was highly effective on most commercial wood species (36,37). A review of the benefits and drawbacks of the use of penta as part of the U.S. Environmental Agency Rebuttable Presumption against Registration Process examined the potential health and environmental impacts of anti-stain treatments (44). This review resulted in the restriction of pentachlorophenate use to those who had passed a test administered by an appropriate state regulatory agency. Concerns about worker safety led many mills to seek alternative chemicals. Evaluation of surface waters around some lumber mills also suggested that run-off of chemicals from freshly treated lumber resulted in high levels of chemical in waters that were crucial for development of many riparian species, notably endangered or threatened salmonids. These concerns led to regulatory actions to ensure that chemicals used for stain prevention did not have adverse aquatic toxicological properties.

The concerns with the safety aspects of the existing chemical treatments have encouraged a continuing search for alternative chemicals and a variety of these materials are now commercially used. Longterm, however, many mills are installing increased kiln capacity to reduce the need for chemicals (26). These activities are occurring despite the much higher costs associated with kiln drying and suggest that many producers are willing to move away from chemicals at almost any cost.

While kiln drying eliminates the need for anti-stain chemicals, non-chemical stain prevention may also be accomplished by the use of competitive or antagonistic organisms that are applied to the wood immediately after sawing to prevent fungal discoloration of wood until the moisture content declines below that required for fungal growth. Biological control has a long history of use in agriculture (2), and the wood surface bears many similarities to the phylloplane (24,27,35,43). The potential for biological control of fungal stain using bacteria or fungi has been extensively explored, often with mixed results (4-12,18,19,21,22,28,34,40,41). In many instances, organisms that are highly effective under controlled laboratory conditions provide inconsistent protection under field regimes. These failures may reflect the range of conditions affecting biocontrol function. These include the presence of other, competing fungi and bacteria on or in the wood at the time of sawing, the nutritional value of the wood, temperature, and moisture content. The performance failures of biocontrol agents create a unique

J. J. Morrell, Department of Forest Products, Oregon State University, Corvallis, Oregon and B. E. Dawson-Andoh, Division of Forestry, West Virginia University Morgantown, West Virginia.

opportunity to identify the factors that most directly affect biocontrol efficiency.

Altering temperature, incubation time of the biocontrol agent prior to wood application, and nutrients in the biocontrol incubation medium can all influence biocontrol efficacy as measured by subsequent degree of wood discoloration (9,28,34). The effects of these variables on the ability of stain fungi to colonize wood and utilize the nutrients present, however, remains less well understood. In order to better understand the problems associated with biocontrol of stain fungi, it would be helpful to review the relationship between each variable and biocontrol performance. This information can then be used to identify the parameters that are most likely to affect biocontrol performance under field conditions.

In reviewing the prior research, it is important to note that studies have been performed on different wood species under different environmental regimes and most importantly, with an array of possible organisms. In addition, it is exceedingly difficult to clearly distinguish which, if any, mechanisms contribute to control in a given system. Furthermore, the opaque nature of wood makes it difficult to observe interactions in situ. Thus, our understanding of biocontrol action in situ evolves from indirect measurements on the substrate such as degree of wood discoloration or enzyme activities or from tests on artificial substrates such as agar. None of these approaches is entirely satisfactory, but they offer insights into potential mechanisms by which these organisms limit microbial discoloration.

The two mechanisms by which biocontrol agents most likely function in stain prevention are competition and antibiosis (3,24). Mycoparasitism might also be considered, but the goal of biocontrol is preventing stain fungi from becoming established and mycoparasitism generally requires the presence of the target fungus (16). It is likely that such establishment would also lead to melanization that would negate the value of any biocontrol. Studies of selected *Trichoderma harzianum* isolates and *Serratia marescens* suggest that chitinase levels produced by these mycoparasites was poorly correlated with biocontrol efficacy (30) (Fig. 1). Chitin stimulates chitinase production by the test organisms, but chitinase levels were not associated with complete inhibition of stain. Chitinase levels in extracts from wood inoculated with both stain fungi and selected *T. harzianum* isolates were poorly correlated with stain inhibition. These results suggest that cell wall degradation is not a primary mechanism by which *T. harzianum* inhibits stain fungi under the conditions tested.

The function of antibiosis in stain prevention by biocontrol agents has been the subject of extensive study for a variety of applications (13,14,23,27). A majority of these studies has been performed using liquid or agar plate cultures. While such studies provide a simple, rapid method for assessing the ability of an organism to produce antibiotics, they are in no way representative of such capabilities on or in wood. The wood substrate is characterized by low levels of nitrogen and other nutrients (17). While there is an abundance of carbohydrate most of this material is not accessible to microorganisms. In solid wood, the organism must grow through this largely inaccessible polymerized mass to obtain nutrients that are concentrated in the ray cells (45). Liquid or agar based growth media are typically rich in nutrients. In addition, the organism is surrounded by these materials in the liquid medium. These differences do not preclude a role for antibiosis in bioprotection, they only suggest the risk in extrapolating laboratory outcomes on artificial substrates to field trials. It is also important to note that

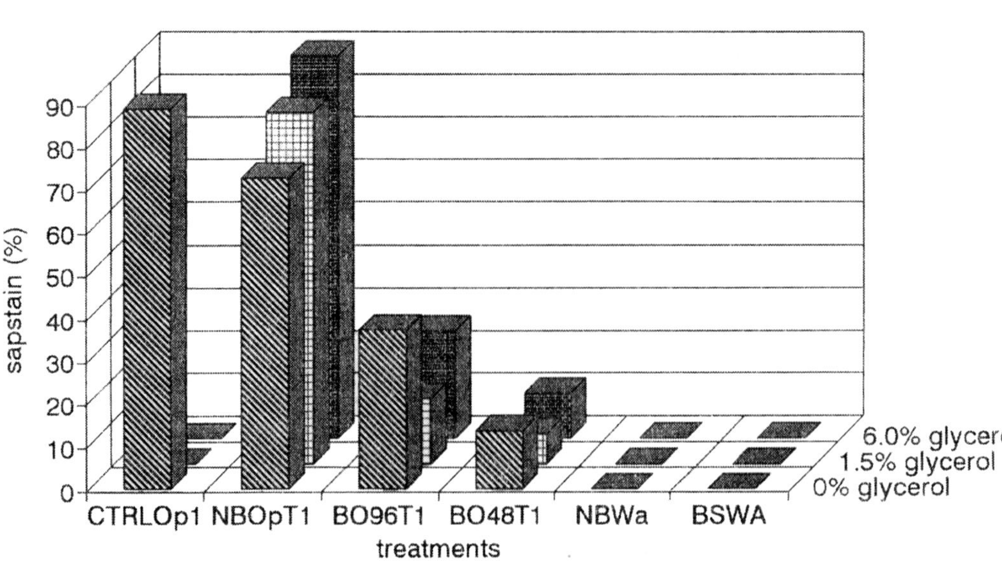

Figure 1.—Degree of discoloration vs biocontrol treatments (Trial 1). Effect of supplemental carbon (glycerol) in the initial bacterial growth media on the ability of *Bacillus subtilis* to inhibit discoloration if ponderosa pine sapwood by *Ophiostoma piliferum*.

relatively small amounts of antibiotic may be capable of inhibition. These levels may defy detection in or on the wood, making it difficult to determine the role of antibiosis in biocontrol. Despite the limitations, a number of antibiotic compounds have been implicated in biocontrol, and it is likely that many more await discovery. The roles of these compounds in biocontrol and the potential for manipulating the environment to enhance their production in situ merits further attention.

In general, competition is considered to be a more important mechanism for biocontrol. Competition can occur for the wood matrix or for various available nutrients (24). Organisms that can rapidly invade wood and sequester any available nutrients have a tremendous advantage over slower growing organisms. Since most stain fungi are capable of rapid growth, potential biocontrol agents must either match this rapid growth rate, or they must be applied at sufficiently high dosages to ensure uniform protection. Confirming the role of competition in biocontrol can be difficult since other possible mechanisms such as antibiosis or mycoparasitism must be shown to be absent. Clearly, competition plays an important role in the effectiveness of several fungal biocontrol agents including *Trichoderma harzianum* and *Gliocladium deliquescens*. These fungi also produce antibiotics in culture, and the former species is capable of mycoparasitism under certain conditions. Delineating the role of each mechanism in successful biocontrol poses a major challenge.

One factor that may affect competitive outcome is the nutritional quality of the wood substrate. Neither the stain fungi or the biocontrol agents are adapted to utilize the lignocellulosic matrix. As a result, these organisms are dependent on their ability to utilize compounds stored in the ray cells. Recent studies indicate that these compounds include lipids, proteins, carbohydrates, and resin acids (1,25). The ability of potential biocontrol agents to colonize the substrate will be highly dependent on the ability to rapidly utilize these compounds ahead of competing microorganisms. This is a potential problem for non-filamentous organisms, particularly bacteria, since they must be capable of moving through the cell matrix to reach the material stored in the parenchyma. As a result, the nutritional surface characteristics of the wood may play a critical role in colonization by such organisms. One approach for aiding colonization is manipulation of the nutrients prior to or during application of the biocontrol organism on the wood surface (31). This can be easily accomplished by adding selected nutrients to the media used to incubate the bioprotectant prior to wood application. However, limited studies using glycerol as a carbon source suggest that nutrient levels had little effect on the degree of fungal stain, nor did added nutrients enhance

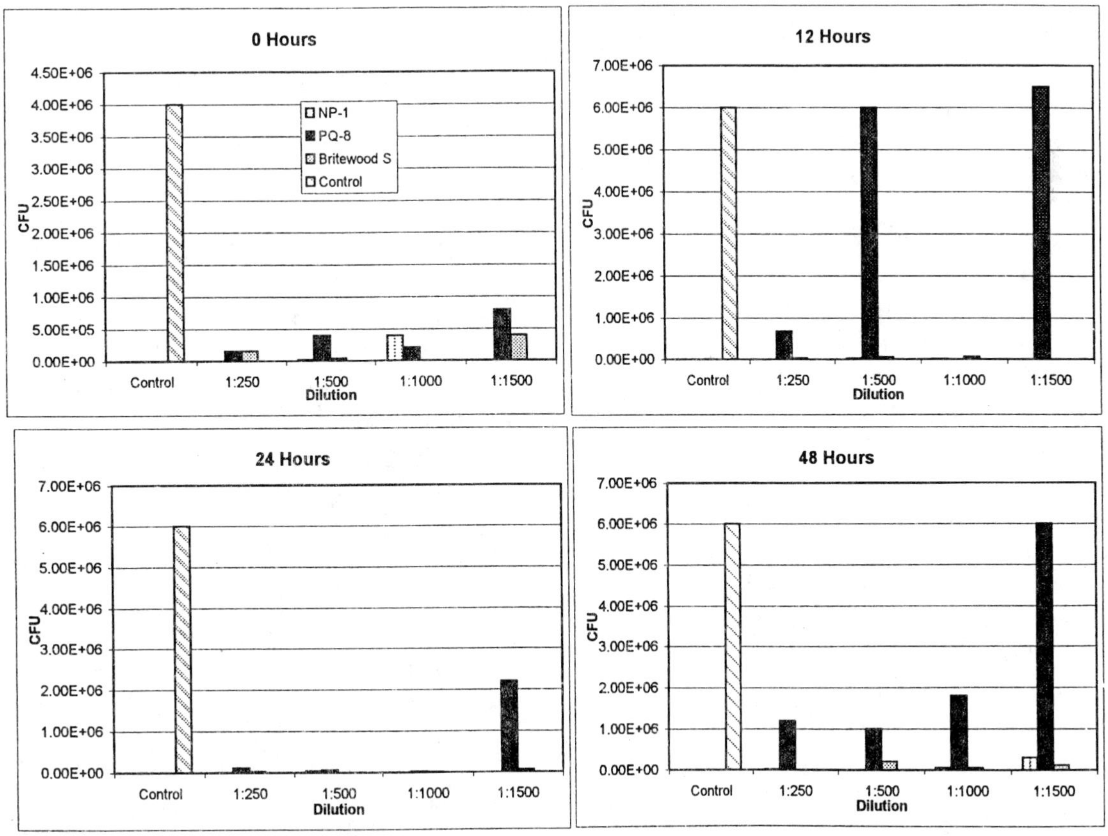

Figure 2.—Number of colony forming units (CFUs) in the presence of selected anti-stain compounds for 0 to 48 hours.

Table 1.—Degree of discoloration of ponderosa pine sapwood wafers 30 days after treatment with *B. subtilis* followed by selected fungicides.

Chemical treatment	Dilution	Degree of discoloration[a] No organisms	Bacterial only	Stain fungus only	Stain fungus/ bacteria
None	–	54 (30)	34 (37)	83 (28)	38 (40)
NP-1	1:250	0 (0)	9 (18)	1 (2)	13 (23)
	1:5000	0 (0)	4 (9)	1 (3)	17 (36)
	1:1000	0 (0)	2 (5)	4 (9)	10 (19)
	1:1500	4 (13)	4 (8)	67 (40)	13 (25)
PQ-8	1:250	2 (5)	2 (6)	3 (6)	15 (19)
	1:500	9 (16)	1 (3)	60 (22)	10 (8)
	1:1000	16 (23)	11 (23)	100 (1)	20 (35)
	1:1500	5 (7)	10 (14)	98 (3)	23 (29)
Britewood S	1:250	7 (9)	17 (22)	82 (16)	5 (9)
	1:500	5 (8)	5 (12)	94 (7)	13 (21)
	1:1000	19 (28)	28 (33)	91 (17)	16 (61)
	1:1500	25 (29)	8 (12)	96 (6)	28 (36)

[a]Values represent means of 15 or 25 samples. Values in parentheses represent one standard deviation.

biocontrol activity (42) (Fig. 1). This does not preclude the use of additives to enhance activity, but it suggests that more thought must be given to nutrient selection. For example, the selection of nutrients that are preferentially used by the biocontrol agent might improve colonization and thus, stain prevention. The primary difficulty in this strategy for enhancing biocontrol is the identification of nutrients that are selectively used by the biocontrol agent. This is particularly difficult when fungal biocontrol organisms are employed. Another disadvantage of this approach is the potential for depletion of the added nutrients during incubation and prior to application to the wood. This problem can be overcome by supplemental nutrient addition immediately before wood application.

An alternative to nutrient enhanced biocontrol is the simultaneous application of low levels of toxins followed by the biocontrol agent (15,20,33). Chemicals for this purpose might include commonly used anti-stain chemicals, albeit at more dilute levels, or other compounds that selectively perturb the target stain fungi, making them more susceptible to the biocontrol organism. These compounds, by their nature, should have minimal toxicity to nontarget organisms. Compounds such as boron and fluoride have been explored for this purpose. Fluoride markedly enhances the growth of *Trichoderma* spp. many of which can inhibit stain and decay fungi (36). Boron has similar effects on some bacteria and simultaneous application of biocontrol agents and mild chemosterilants may enhance biocontrol activity. Similarly, it may be possible to treat wood with a biocontrol immediately after cutting or sawing followed by a dilute solution of fungicide. Trials with a number of commercially used anti-stain chemicals and a biocontrol organism, *Bacillus subtilis* suggest that fungal stain can be inhibited to a greater extent when combinations of biocide and biocontrol are employed (32) (Table 1). This effect was negligible when the biocontrol was added to the anti-stain chemical and applied to the wood. The lack of benefit reflected the devastating effect of the biocide on bacterial fitness. The number of colony forming units in the bacterial inoculum declined precipitously immediately after mixing with the biocide (Fig. 2). Separate applications, however, reduced this reaction and resulted in more effective control than was found with either the biocides or the bioprotectant alone. The strategy of simultaneous application is probably not practical in a mill situation, but it may be useful for protecting wood between the time of felling and sawing. The biocontrol agent could be sprayed or brushed on the cut surfaces of logs in the woods with minimal environmental risk. This application would protect the logs from stain until they could be processed. A more conventional biocide would then be used, albeit at a lower level to protect the wood until it was dried below 20 percent moisture content. Clearly, the applications for this approach are minimal, but the use of biocontrols on a limited scale where maximum environmental control can be exerted, may represent a sound initial approach to developing these strategies for more robust applications.

Conclusion

It is clear that biocontrol of stain fungi on the surfaces of freshly sawn wood has tremendous potential. At present, however, the state of knowledge related to the effects of environmental variables on the performance of biocontrol agents on wood surface is limited. In the absence of such knowledge, biocontrol strategies will continue to provide inconsistent protection of negligible value for commercial prevention of fungal stain.

Literature Cited

1. Abraham, L.D. and C. Breuil. 1993. Organic nitrogen in wood: Wood substrates for a sapstain fungus. IRG/WP/10019.
2. Baker, C.J. and R.J. Cook. 1987. Evolving concepts of biological control of plant pathogens. Ann. Rev. Phytopath. 25:67-85.
3. Barnett, H.L. and F.L. Binder. 1973. The fungal host-parasite relationship. Ann. Rev. Phytopath. 11:273-292.
4. Benko, R. 1986. Protection of wood against the blue stain. Int. Congress of IUFRO, Ref. #18, Ljubljana, Yugoslavia.

5. Benko, R. 1987. Antagonistic effects of some mycorrhizal fungi as biological control of blue-stain. Int. Res. Group on Wood Preserv. IRG/WP/1314. Stockholm, Sweden.
6. Benko, R. 1988. Bacteria as possible organisms for biological control of blue-stain. Int. Res. Group on Wood Preserv. IRG/WP/1339. Stockholm, Sweden.
7. Benko, R. 1989. Bacterial control of blue-stain on wood with *Pseudomonas cepacia* 6253. Laboratory and field test. Int. Res. Group on Wood Preserv. IRG/WP/1380. Stockholm, Sweden.
8. Benko, R. and B. Henningsson. 1986. Mycoparasitism by some white rot fungi on blue stain in culture. Int. Res. Group on Wood Preserv. IRG/WP/1304. Stockholm, Sweden.
9. Benko, R. and T.L. Highley. 1990a. Selection of media on screening interaction of wood attacking fungi and antagonistic bacteria. 1. Interaction on agar. Mat. und Org. 25(3):161-171.
10. Benko, R. and T.L. Highley. 1990b. Selection of media on screening interaction of wood attacking fungi and antagonistic bacteria. 2. Interaction on wood. Mat. und Org. 259(3):173-180.
11. Bernier, R., J.M. Desrochers and L. Jursek. 1986. Antagonistic effect between *Bacillus subtilis* and wood staining fungi. J. Inst. Wood Science, 10(5):214-216.
12. Blanchette, R.A., R.L. Farrell, T.A. Burnes, P.A. Wendler, W. Zimmerman, T.S. Brush, and R.A. Snyder. 1992. Biological control of pitch in pulp and paper production by *Ophiostoma piliferum*. Tappi 77(1):155-159.
13. Bruce, A. and B. King. 1983. Biological control of wood decay by Lentinus lepideus (Fr.) produced by *Scytalidium* and *Trichoderma residues*. Mat. und Org. 18:171-181.
14. Bruce, A., W.J. Austin and B. King. 1984. Control of growth of *Lentinus lepideus* by volatiles from Trichoderma. Trans. Brit. Mycol. Soc. 82:423-428.
15. Chet, I.A. 1990. Biological control of soil-borne pathogens with fungal antagonists in combination with soil treatments. In: Biological Control of Soil-Borne Plant Pathogens (D. Hornby ed.), CAB International, Wallingford, U.K. pp. 15-25.
16. Chet, I. 1990. Mycoparasitism - recognition, physiology and ecology. In: New Directions in Biological Control: Alternatives for Suppressing Agricultural Pests and Diseases. Alan L. Liss, Inc. pp. 725-733.
17. Cowling, E.B. and W. Merrill. 1966. Nitrogen in wood and its role in wood deterioration. Can. J. Bot. 44:1,539-1,554.
18. Croan, S.C. and T.L. Highley. 1991. Control of sapwood-inhabiting fungi by fractionated extracellular metabolites from *Coniophora puteana*. Int. Res. Group Wood Preserv. IRG/WP/1494. Stockholm, Sweden.
19. Croan, S.C. and T.L. Highley. 1992. Biological control of sapwood-inhabiting fungi by living bacterial cells of *Streptomyces rimosus* as a bioprotectant. Int. Res. Group on Wood Preserv. IRG/WP/1564-92. Stockholm, Sweden.
20. Dawson-Andoh, B. and J.J. Morrell, 1990. Effects of chemical pretreatment of Douglas-fir heartwood on efficacy of potential bioprotection agents. Int. Res. Group on Wood Preserv. IRG/WP/1440. Stockholm, Sweden.
21. Dawson-Andoh, B.E., M. Chan, B. McAfee, and R. Lovell. 1994. Extracellular enzymes produced by potential bioprotectants and sapstain fungi during colonization of western hemlock. Technical Forum. FPS 48th Annual Meeting, Portland, Maine.
22. Florence, J.M. and J.K. Sharma, 1990. *Botryodiplodia theobromae* associated with blue staining in commercially important timbers of Kerala and its possible biological control. Mat. und Org. 25(3):193-199.
23. Fravel, D.R. 1988. Role of antibiosis in the biocontrol of plant disease. Annual Review of Phytopathology, 26:75-91.
24. Freitag, M., J.J. Morrell, and A. Bruce. 1991. Biological protection of wood: Status and prospects. Biodeterioration Abstracts 5(1):1-13.
25. Gao, Y. and C. Breuil. 1995. Wood extractives as carbon sources for staining fungi in the sapwood of lodgepole pine and trembling aspen. IRG/WP/10098
26. Hansen, E. and J.J. Morrell. 1997. Use of antistain chemical treatments by the western softwood lumber industry: 1994. Forest Products Journal 47(6):69-73.
27. Harman, G.E. 1990. Deployment tactics for biocontrol agents in plant pathology. In: New Directions in Biological Control: Alternatives for Suppressing Agricultural Pests and Diseases. Alan R. Liss. pp. 779-792
28. Kreber, B. and J.J. Morrell. 1993. Ability of selected bacterial and fungal bioprotectants to limit fungal stain in Ponderosa pine sapwood. Wood and Fiber Science 25(1):23-34.
29. Land, C.J., Bandhidi, Z.G. and A.-C. Albertson. 1985. Surface discoloring and blue staining by cold-tolerant filamentous fungi on outdoor softwood in Sweden. Mat. und Org. 20(2):133-156.
30. Liu, J. and J.J. Morrell. Effect of bioconrol inoculum growth conditions on subsequent chitinase and protease levels in wood exposed to biocontrols and stain fungi. Mat. und Org. (in review).
31. Mandelsohn, M.L., T.M. Ellwanger, R.L. Rose, J.L. Kough, and P.O. Hutton. 1995. Registration of biologicals. How product formulations affect data requirement. In: Biorational Pest Control Agents. Formulation and Delivery. F. R. Hall and J.W. Barry (eds.), ACS Symposium Series 595, pp. 20-26.
32. Mankowski, M., M. Anderson, and J.J. Morrell. 1997. Integrated protection of freshly sawn lumber using *Bacillus subtilus* and selected fungicides. International Research Group on Wood Preservation. Document No. IRG/WP/97-70235. Stockholm, Sweden, 7 pages.
33. Morrell, J.J. and C.M. Sexton. 1990. Effects of volatile chemicals on the ability of microfungi to arrest basidiomycetous decay. Mat. und Org. 25(4):267-274.
34. Morrell, J.J. and C.M. Sexton, 1990. Fungal staining of Ponderosa pine sapwood: Effect of preconditioning and bioprotectants. Wood and Fiber Science 25(4):322-325. Int. Res. Group on Wood Preserv. IRG/WP/1551-92. Stockholm, Sweden.
35. Morris, C.E. and D.J. Rouse. 1985. Role of nutrients in regulating epiphytic bacterial populations. In: Biological control on the phylloplane (C.E. Windels and S.E. Lindow eds.). American Phytopathological Society pp. 63-82.

36. Panek, E. 1963. Pretreatments for the protection of southern yellow pine poles during air-seasoning. Proceedings American Wood Preserver's Association, 59:189-202.
37. Rao, K.R. (ed.) 1978. Pentachlorophenol: Chemistry, pharmacology, and environmental toxicology. Environmental Science Research, Vol. 12. Plenum Press, New York, N.Y. 402 pp.
38. Scheffer, T.C. and R.M. Lindgren. 1940. Stains of sapwood products and their control. Techn. Bull. 714. U.S. Dept. of Agric. Washington, D.C.
39. Schaefer, T.C. 1973. Microbial degradation and the causal organisms. In: D.D. Nicholas (ed.). Wood Deterioration and its Prevention by Preservation Treatments. Syracuse Univ. Press, Syracuse, N.Y. pp. 31-106.
40. Seifert, K.A., W.E. Hamilton, C. Breuil, and M. Best. 1987. Evaluation of *Bacillus subtilis* C 186 as a potential biological control of sapstain on mold on unseasoned lumber. Can. J. Microbial. 33:1,102-1,107.
41. Silva, A. and J.J. Morrell. Effects of nutrients on the ability of *B. subtilis* to inhibit staining and enzxyme activity by *O. perfectum* on Ponderosa pine sapwood. Phytopathology (in review).
41. Seifert, K.A., C. Breuil, L. Rossignol, M. Best, and J.N. Saddler. 1988. Screening for microorganisms with the potential for biological control of sapstain on unseasoned lumber. Mat. und Org. 23:81-95.
43. Spurr, H.W. 1990. The phylloplane. IN: New directions in biological control. Alternatives to suppressing agricultural pests and diseases (R.R. Baker and P.E. Dunn eds.). Alan R. Liss Inc., New York, pp. 271-278.
44. U.S. Department of Agriculture. 1980. The biological and economic assessment of pentachlorophenol, inorganic arsenicals, and creosote. Vol. I. Wood Preservatives, USDA Techn. Bull., No. 1658-1. Washington, D.C. 435 pp.
45. Zabel, R.A. and J.J. Morrell. 1992. Wood Microbiology: Decay and its Prevention. Academic Press, Inc., San Diego, CA 474 pp.

A New Approach for Potential Integrated Control of Wood Sapstain

D. Q. Yang

Abstract

Gliocladium roseum has been identified as a potential biological control agent for preventing wood sapstain. This fungus significantly reduced sapstain on pasteurized hemlock lumber; however, limited success has been achieved on non-pasteurized wood. Recently, the effect of pH on growth of sapstaining fungi and *G. roseum* was investigated. Forty-one fungal isolates from 15 species were tested at pH values ranging from 2.4 to 11.2. Results showed that most sapstaining fungi grew well on malt extract agar over a range of pH from 3.5 to 8.9. However, the potential biocontrol agent, *G. roseum*, was extremely tolerant to alkaline conditions. It grew well even on the medium at pH 10.8. A further experiment was carried out to test the efficacy of different alkaline solutions, the biocontrol agent, and the combination of both chemical and biological agents on inhibiting growth of different sapstaining fungi. Results showed that the alkali or the biocontrol agent alone only partially inhibited the growth of sapstaining fungi. However, when using a combination of a suitable alkaline solution and the biocontrol agent to treat the medium before sapstaining fungal spores were added to it, *G. roseum* rapidly occupied the whole plate, whereas growth of sapstaining fungi was suppressed. A similar result was also achieved on wood wafers of white pine in the laboratory condition. These results suggest that simultaneously applying a mild alkali with *G. roseum* may improve the efficiency of *G. roseum* against wood sapstaining fungi on unseasoned lumber.

Introduction

Lumber is subjected to degrade from moulds and sapstaining fungi during storage and shipment. Customers place a high value on the appearance and quality of the product they receive. Wood stain can significantly reduce the value of the lumber by lowering its grade which leads to monetary and market losses. To offset fungal discoloration, lumber is kiln dried or chemically treated after processing in sawmills. Public concern for the environment has encouraged an interest in exploring new technology for wood protection and reducing chemical usage. Biological protection of wood from fungal degradation has been recognized as a possible alternative to chemical treatment and has received intensive investigation worldwide for many years (1,4,13). However, this approach has worked well in laboratory tests for many microorganisms, but there has been limited field success (8,10).

Gliocladium roseum has been identified as a potential biocontrol agent against sapstain on softwood (12). This antagonist protected pasteurized lumber from stain but showed inconsistent performance on green lumber, which may due to the lack of a comprehensive competitive ability against other wood-inhabiting fungi (7). To improve the effectiveness of a bioprotectant, one approach could be to selectively alter conditions of wood with a view to giving the promoting fungus a competitive advantage (2,9). Manipulation of the pH value of wood may be one of such condition that needs to be investigated.

This project was designed to obtain information on pH ranges for growth of moulds and sapstaining fungi and to investigate the feasibility of developing a protective system against wood sapstain based on a combination of an alkaline solution and a potential biocontrol agent.

Materials and Methods

Sources of Organisms

All fungal cultures used in the tests originated from the Forintek Culture Collection maintained on 2 percent malt extract agar (20 g malt extract and 20 g Difco agar

D. Q. Yang, Research Scientist, Forintek Canada Corporation, Quebec, Canada.

in 1 liter distilled water) at 5°C. Cultures were transferred onto the same freshly prepared medium in Petri plates and incubated at 26°C for 1 or 2 weeks before being used in the experiments.

Effect of pH on Spore Germination

Eight moulds and sapstaining fungi were selected for this experiment. Fungi were subcultured on Petri plates containing 2 percent malt extract agar. The plates were incubated at 26°C with continuous light for 14 days. Conidia formed were collected by adding 2 ml of sterile distilled water to each plate and gently agitating it with a glass rod. The suspension was filtered through four layers of muslin, and the final concentration of spores was adjusted to 107 spores/ml with sterile distilled water. A volume of 0.9 ml of filter-sterilized (Millipore, 0.2 µm) 2 percent malt extract broth adjusted to pH 3, 6, 9, or 11 by addition of either 19.6 percent lactic acid or 1N NaOH was filled in each of 1.5 ml micro-tubes, and 0.1 ml of spore suspension was added to each tube to give a spore concentration of 10^6 spores/ml. The tubes were incubated at room temperature (20° ± 2°C), and the spore germination rate was recorded as the average of three replicates from each fungal isolate at 24 and 48 hours after inoculation.

Effect of pH on Mycelial Growth

Forty-one isolates representing 15 moulds and sapstaining fungi were selected for this experiment (Table 2). Cultures were grown on Petri plates containing 2 percent malt extract agar and incubated at 26°C for 14 days. Fresh spore suspensions were prepared by washing each plate with 2 ml of sterile distilled water and gently agitating it with a glass rod. The suspension was filtered through four layers of muslin, and the spore concentration was adjusted to 10^6 spores/ml water. A modification of the method of Dennis (3) was used to adjust pH value of the medium. One liter of 2 percent malt extract agar in a 2-liter flask was autoclaved for 15 minutes at 120°C, cooled to 45°C, and adjusted to different target pH levels from 2.5 to 11.5 by adding pre-measured required amounts of filter-sterilized either 19.6 percent lactic acid or 1N NaOH. Accurate pH values were determined by measuring a portion of the medium in each flask with a pH meter before pouring it into 9-cm diameter Petri plates. A volume of 0.2 ml of each spore suspension was applied onto each plate and was evenly spread over the entire surface with a glass rod. The plates were incubated at 26°C for 7 days, after which amount of mycelial growth was rated on a scale of 0 to 5 as follows:

0 = no growth;
1 = trace of growth;
2 = growth covered less than 25% of the surface;
3 = moderate growth on the surface with 25% to 50% coverage;
4 = heavy growth on the surface with more than 50% and less than 75% coverage;
5 = very heavy growth on the surface with more than 75% coverage.

Assessment was based on replicate plates in each treatment from 2 separate tests.

Survival of Fungi on Malt Extract Agar Supplemented with an Alkali and a Biocontrol Agent

Based on results obtained from the test described above, five alkali-tolerant sapstaining fungi and one isolate of *Gliocladium roseum* (321M) were selected for this experiment. Petri plates (5.5 cm in diameter) containing 10 ml of 2 percent malt extract agar were prepared, and each plate was evenly applied over the entire surface with 0.4 ml of one of the following solutions:

1. sterile distilled water;
2. *G. roseum* spore suspension at 1×10^6 spores/ml in water;
3. 5 percent sodium carbonate solution in water; and
4. *G. roseum* spore suspension at 1×10^6 spores/ml in 5 percent sodium carbonate solution.

After the treatment, the Petri plates were divided into two groups. The first group was designed to test the competitive ability of *G. roseum* against sapstaining fungi when they presented in the same conditions. In this group, lids of the plates were left open until the medium had completely absorbed the treating solutions approximately 30 minutes later, then 0.2 ml of each of the sapstaining fungal spore suspension was evenly applied over the surface of the plates, respectively. The second group was designed to test the protection ability of *G. roseum* against staining fungi after it first established on the host. In this test, the treated plates were first incubated at 26°C for 48 hours, then 0.2 ml of each of sapstaining spore suspensions was evenly applied over the plates. In all cases, three plates (replications) were used. All plates were incubated in a growth chamber at 26°C and 75 percent relative humidity, and growth of sapstaining fungi in the plates was assessed after 4 weeks of incubation based on a scale of 0 to 5.

Wood Wafer Tests

Freshly cut sugar maple (*Acer saccharum*) and white pine (*Pinus strobus*) bolts were obtained from local sawmills in Quebec and kept at –30°C in a cold room. Prior to the test, the wood bolts were thawed and wood wafers (4 by 2 by 0.5 cm) were cut from defect-free green sapwood of the bolts. Spore suspension (1×10^6

spores/ml) was made from isolate 321M of *G. roseum* by the method described above. A mixed spore suspension (1×10^6 spores/ml) was also prepared in the same way as with *G. roseum* using equal amount of spores from the following 10 sapstaining fungi: *Ophiostoma piceae* 387E, *Aureobasidium pullulans* 132S, *Ceratocystis adiposa* 251E, *Alternaria alternata* 2G, *Cladosporium sphaerospermum* 806B, *Hormonema dematioides* 742D, *Rhinocladiella atrovirens* 135E, *Ophiostoma minus* 864A, *Phialophora botulispora* 707C, and *Ophiostoma piliferum* 55B.

Wood wafers were dipped for 30 seconds in the following solutions:

1. sterile distilled water;
2. *G. roseum* spore suspension at 1×10^6 spores/ml in water;
3. 5 percent sodium carbonate solution in water; and
4. *G. roseum* spore suspension at 1×10^6 spores/ml in 5 percent sodium carbonate solution.

After the treatment, wood wafers were divided into two groups. In the first group, treated samples, six pieces per treatment, were immediately inoculated with 0.1 ml of the mixed sapstaining spore suspension and were placed, two pieces per plate, on a W-shaped glass supporter sitting on two layers of wet filter paper in a Petri plate. In the second group, six samples, in each treatment, were placed in three Petri plates in the same way as with the samples in the first group. However, these treated wafers were incubated at 26°C for 3 days before they were inoculated with 0.1 ml of a mixed sapstaining spore suspension. All samples were incubated in an environmentally controlled chamber set at 26°C and 75 percent relative humidity, and they were inspected after 8 weeks of incubation on a scale of 0 to 5. For each treatment, three aspects of information were provided: 1) average score, which measures the general severity of the infection, was obtained by averaging stain ratings from all pieces in a treatment; 2) clean pieces which indicates the number of uninfected wood pieces in a treatment; and 3) acceptable pieces which demonstrates the number of pieces that have a score of 2 or less which is considered to be commercially acceptable on a 2 by 4 construction commodity in global markets.

Statistical Analysis

Data were subjected to analysis of variance (ANOVA) using the statistical analysis system (11). Following the ANOVA, the individual means were compared using Scheffe's test for multiple comparison.

Results and Discussion

Effect of pH on Spore Germination

As shown in Table 1, the optimum pH for spore germination of sapstaining fungi was 3 to 6, while it was 6 to 9 for *Gliocladium roseum*. At pH 9, spore germination from all fungi, except *Hormonema dematioides* at 48 hours, was significantly reduced. At pH 11, no germination was found from *Ophiostoma piceae*, *Leptographium* sp., and *Trichoderma harzianum*, and very low germina-

Table 1.—Effect of pH on spore germination of moulds and sapstaining fungi.[a]

Test fungus	Incubation period (hours)	Average germination (%)			
		pH 3	pH 6	pH 9	pH 11
Ophiostoma piceae 387E	24	18.55 b	57.86 a	6.41 c	0.00 c
	48	64.11 b	76.61 a	6.17 c	0.00 c
Ophiostoma piliferum 55F	24	25.40 b	50.02 a	4.93 c	0.00 c
	48	64.14 a	67.07 a	19.59 b	3.17 c
Leptographium sp. 865A	24	26.86 a	35.14 a	2.21 b	0.00 b
	48	27.11 b	47.06 a	7.18 c	0.00 c
Hormonema dematioides 742D	24	6.29 ab	9.16 a	4.22 bc	1.06 c
	48	11.48 b	22.34 a	17.67 ab	0.86 c
Aureobasidium pullulans 132S	24	1.59 b	6.58 a	1.52 b	0.00 b
	48	2.24 b	5.30 a	2.21 b	1.42 b
Rhinocladiella atrovirens 135E	24	6.93 ab	12.90 a	4.24 bc	0.00 c
	48	10.12 b	19.00 a	8.16 bc	0.67 c
Trichoderma harzianum 160C	24	80.03 a	36.82 b	2.44 c	0.00 c
	48	82.40 a	42.17 b	7.14 c	0.00 c
Gliocladium roseum 321U	24	52.13 c	82.87 a	73.98 ab	60.21 bc
	48	61.79 c	91.45 a	95.53 a	76.63 b

[a] Values are means of three replicates. Means followed by the same letters in a row are not significantly ($p = 0.05$) different from each other by Scheffe's test.

Table 2.—Growth rating[a] of moulds and sapstaining fungi on malt extract agar at different pH values[b].

Fungus (isolate)	pH 2.4 (2.5)	pH 2.9 (3.0)	pH 3.5 (3.5)	pH 5.5 (6.0)	pH 8.9 (9.0)	pH 9.6 (10.0)	pH 10.8 (11.0)	pH 11.2 (11.5)
Gliocladium roseum								
321A	0.0 (0.0)	1.0 (0.0)	3.0 (0.0)	4.3 (0.5)	5.0 (0.0)	4.5 (0.6)	2.5 (1.0)	2.8 (0.5)
321H	0.0 (0.0)	1.5 (0.6)	4.0 (0.0)	5.0 (0.0)	4.8 (0.0)	4.5 (0.6)	4.5 (0.6)	3.0 (0.0)
321M	0.0 (0.0)	1.0 (0.0)	5.0 (0.0)	5.0 (0.0)	5.0 (0.0)	4.5 (0.6)	4.0 (0.0)	3.0 (0.0)
321U	0.0 (0.0)	1.5 (0.6)	4.0 (0.0)	5.0 (0.0)	5.0 (0.0)	4.5 (0.6)	4.5 (0.6)	3.5 (0.6)
Ophiostoma piceae								
387D	0.0 (0.0)	1.0 (0.0)	4.0 (0.0)	4.8 (0.5)	4.5 (0.6)	3.5 (0.6)	1.0 (0.8)	0.0 (0.0)
387E	0.0 (0.0)	2.0 (1.2)	4.0 (0.0)	4.5 (0.6)	4.5 (0.6)	4.5 (0.6)	3.8 (0.5)	0.0 (0.0)
387I	0.0 (0.0)	1.5 (0.6)	5.0 (0.0)	4.5 (0.0)	4.0 (0.0)	3.8 (1.0)	1.5 (1.3)	0.0 (0.0)
Aureobasidium pullulans								
132B	0.5 (0.6)	2.0 (0.0)	5.0 (0.0)	4.5 (0.6)	3.5 (0.6)	3.5 (0.6)	0.0 (0.0)	0.0 (0.0)
132Q	0.5 (0.6)	3.0 (0.0)	5.0 (0.0)	4.8 (0.5)	4.8 (0.5)	1.8 (0.5)	0.0 (0.0)	0.0 (0.0)
132S	1.0 (0.0)	3.8 (1.0)	5.0 (0.0)	4.0 (0.0)	3.5 (0.6)	3.3 (0.5)	2.0 (0.0)	0.8 (0.5)
Leptographium sp.								
865A	0.0 (0.0)	1.5 (0.6)	4.0 (0.0)	4.5 (0.6)	4.0 (0.0)	4.0 (0.0)	0.0 (0.0)	0.0 (0.0)
Ceratocystis adiposa								
251A	0.0 (0.0)	2.5 (0.6)	5.0 (0.0)	4.8 (0.5)	5.0 (0.0)	4.5 (0.6)	3.3 (0.5)	0.5 (1.0)
251B	0.0 (0.0)	2.5 (0.6)	5.0 (0.0)	4.5 (0.6)	4.5 (0.6)	3.8 (0.5)	2.8 (0.5)	0.0 (0.0)
251E	0.0 (0.0)	2.5 (0.6)	5.0 (0.0)	4.3 (0.5)	5.0 (0.0)	4.5 (0.6)	3.5 (0.6)	1.0 (1.2)
Hormonema dematioides								
742A	0.0 (0.0)	1.5 (0.6)	5.0 (0.0)	4.3 (0.5)	4.0 (0.0)	3.5 (0.6)	2.5 (0.6)	0.0 (0.0)
742C	0.0 (0.0)	1.5 (0.6)	5.0 (0.0)	4.3 (0.5)	4.5 (0.6)	4.0 (0.0)	2.8 (0.5)	1.0 (0.8)
742D	0.0 (0.0)	2.5 (1.0)	5.0 (0.0)	4.5 (0.6)	4.5 (0.6)	3.3 (1.0)	2.3 (1.0)	0.0 (0.0)
Rhinocladiella atrovirens								
135E	0.0 (0.0)	3.5 (0.6)	5.0 (0.0)	4.0 (0.0)	3.5 (0.6)	3.3 (1.0)	4.5 (0.6)	0.0 (0.0)
135I	0.0 (0.0)	4.0 (0.0)	5.0 (0.0)	4.5 (0.6)	4.0 (0.0)	3.5 (0.6)	1.5 (0.6)	0.0 (0.0)
135M	0.0 (0.0)	3.0 (1.2)	5.0 (0.0)	5.0 (0.0)	4.8 (0.5)	3.5 (1.7)	1.0 (0.8)	0.0 (0.0)
Ophiostoma minus								
864A	0.0 (0.0)	1.0 (0.0)	5.0 (0.0)	5.0 (0.0)	4.5 (0.6)	3.8 (0.5)	0.0 (0.0)	0.0 (0.0)
Sporothrix sp.								
718B	2.0 (1.2)	2.5 (1.7)	4.5 (0.6)	3.8 (0.5)	3.0 (0.0)	1.5 (0.6)	1.0 (0.0)	0.0 (0.0)
718D	2.0 (1.2)	2.5 (1.7)	4.0 (0.0)	3.5 (0.6)	2.5 (0.6)	1.5 (0.6)	0.0 (0.0)	0.0 (0.0)
718F	2.0 (1.2)	2.5 (1.7)	4.0 (0.0)	3.5 (0.6)	2.5 (0.6)	1.5 (0.6)	0.0 (0.0)	0.0 (0.0)
Phialophora botulispora								
707A	0.0 (0.0)	0.5 (0.6)	2.0 (1.2)	4.5 (0.6)	4.5 (0.6)	3.8 (0.5)	2.3 (1.0)	0.8 (0.5)
707B	0.0 (0.0)	3.0 (0.0)	5.0 (0.0)	4.5 (0.6)	5.0 (0.0)	1.5 (0.6)	0.0 (0.0)	0.0 (0.0)
707C	0.0 (0.0)	3.5 (0.6)	5.0 (0.0)	4.0 (0.6)	3.8 (0.5)	2.5 (0.6)	0.0 (0.0)	0.0 (0.0)
Alternaria alternata								
2A	0.0 (0.0)	0.5 (0.6)	3.0 (0.0)	4.5 (0.6)	4.5 (0.6)	3.0 (1.2)	3.0 (1.2)	2.0 (1.2)
2G	0.0 (0.0)	1.0 (0.0)	2.5 (0.6)	4.5 (0.6)	4.5 (1.0)	4.5 (0.6)	4.0 (0.0)	2.0 (1.2)
2H	0.0 (0.0)	0.3 (0.5)	2.0 (1.2)	4.5 (0.6)	4.5 (0.6)	3.5 (0.6)	2.5 (0.6)	1.3 (1.0)
Ophiostoma piliferum								
55B	0.0 (0.0)	0.5 (0.6)	5.0 (0.0)	4.5 (0.6)	4.0 (0.0)	3.0 (1.2)	0.0 (0.0)	0.0 (0.0)
55F	0.0 (0.0)	2.0 (1.2)	4.0 (0.0)	4.8 (0.5)	4.3 (1.0)	4.3 (1.0)	0.0 (0.0)	0.0 (0.0)
55H	0.0 (0.0)	2.0 (0.0)	5.0 (0.0)	4.3 (0.5)	4.0 (0.0)	2.5 (1.7)	0.0 (0.0)	0.0 (0.0)
Ophiostoma sagmatospora								
832B	0.0 (0.0)	1.5 (0.6)	4.0 (0.0)	5.0 (0.0)	3.5 (0.6)	3.3 (1.0)	0.0 (0.0)	0.0 (0.0)
832C	0.0 (0.0)	2.0 (1.2)	4.0 (0.0)	5.0 (0.0)	4.5 (1.0)	3.8 (1.0)	0.0 (0.0)	0.0 (0.0)
832D	0.0 (0.0)	1.5 (0.6)	4.0 (0.0)	5.0 (0.0)	4.0 (1.4)	4.0 (1.2)	0.0 (0.0)	0.0 (0.0)
Cladosporium sphaerospermum								
806A	0.0 (0.0)	1.0 (1.2)	3.5 (0.6)	5.0 (0.0)	5.0 (0.0)	4.5 (0.6)	3.5 (0.6)	2.5 (0.6)
806B	0.5 (0.6)	1.5 (0.6)	4.0 (0.0)	5.0 (0.0)	4.8 (0.5)	4.3 (1.0)	4.0 (1.2)	3.0 (0.6)
Graphium sp.								
664I	0.0 (0.0)	1.0 (1.2)	3.5 (0.6)	4.5 (0.6)	4.8 (0.5)	2.5 (1.7)	0.5 (0.6)	0.0 (0.0)
664R	0.0 (0.0)	1.5 (0.6)	5.0 (0.0)	4.0 (0.0)	4.5 (0.6)	4.0 (1.2)	0.0 (0.0)	0.0 (0.0)
664aL	0.0 (0.0)	2.5 (0.6)	5.0 (0.0)	4.5 (0.6)	5.0 (0.0)	4.0 (1.2)	0.0 (0.0)	0.0 (0.0)

[a] Data are means of four replicates from two separate experiments with standard deviation in parentheses.
[b] Target values are presented in parentheses below accurate values.

tion was observed from the rest of the staining fungi, except that *G. roseum* still showed a high germination rate at this pH point. This indicates that changing the wood surface pH value to over 6 may reduce initial infection from sapstaining fungi and enhance the colonization of the biocontrol agent.

Effect of pH on Mycelial Growth on Solid Medium

Variation in growth between and within species of sapstaining fungi response to pH *in vitro* is presented in Table 2. In general, it is evident that most testing fungi had a fairly large pH tolerance range between pH 2.9 and 10.8, with a few isolates that survived at pH 2.4 or 11.2. Though seven isolates from three species survived at pH 2.4, only *Sporothrix* sp. showed fairly good growth. At pH 11.2, trace growth was found from 10 of 37 sapstaining fungi investigated, but only isolates from *Alternaria alternata* and *Cladosporium sphaerospermum* developed appreciably. All isolates of *G. roseum* still grew fairly well at pH 11.2.

The comparison of different isolates within a species revealed significant differences in response to pH. Taking *Aureobasidium pullulans* for example, growth of isolate 132B started at pH 2.4, peaked at pH 3.5 to 5.5, declined with further pH increase, and stopped at pH 10.8. Isolate 132Q showed a pattern similar to 132B, but attained its highest growth level from pH 3.5 to 8.9. Isolate 132S also started its growth at pH 2.4, but peaked rather strikingly at pH 3.5, declined gradually with increase of pH, and still grew at pH 11.2. Similar examples were also found within other species investigated in this experiment and elsewhere (5,6).

This work has shown that moulds and sapstaining fungi can grow and develop in a wide pH range and that a pH of 9.6, in most cases, reduces but does not prevent growth. The suggestion of changing wood pH by treating the wood surface with alkaline solution alone, in order to prevent growth of these fungi, does not seem to be a realistic practice. The wood surface is difficult to keep at a higher pH than 9.6 for long time because of the buffering capacity of the wood and the fade out of the applied solution with time. However, using an alkaline solution to suppress sapstaining fungi and to help *G. roseum* establish itself quickly on wood seems to be realistic.

Survival of Fungi on Malt Extract Agar Supplemented with an Alkali and a Biocontrol Agent

Results on immediate-application of individual sapstaining fungus on malt extract agar following different treatment are shown in Table 3. In the absence of sodium carbonate, *G. roseum* only partially restricted growth from *O. piceae, A. pullulans, A. alternata,* and *C. spherospermum*, but growth of *Ceratocystis adiposa* was not reduced. While in the presence of sodium carbonate, *G. roseum* grew rapidly all over the plate and four of five testing fungi, except *A. alternata*, were totally eliminated. Table 3 also presents results of post-application of sapstaining fungi on malt extract agar following the treatment. In the absence of sodium carbonate, though *G. roseum* had established well on the plates, only *C. adiposa* was totally eliminated. *O. piceae, A. pullulans, A. alternata,* and *C. spherospermum* were still able to grow more or less on the plates. However, in the presence of 5 percent sodium carbonate, growth of *O. piceae, A. pullulans, C. adiposa,* and *C. spherospermum* was totally eliminated, while that of *A. alternata* was restricted to a scale below 0.4.

Comparing results from the test of immediate-inoculation with that of post-inoculation indicates that *G. roseum* is unable to compete successfully with all sapstaining fungi on malt agar without the presence of sodium carbonate. Certain species, such as *C. spherospermum*, grew fairly well even when *G. roseum* was first well established. However, the competitive and antagonistic

Table 3.—Growth of sapstaining fungi on malt extract agar following immediate exposure and post-exposure to different testing agents in the plates.[a]

Treatment	Average stain rating: Immediate exposure	Average stain rating: Post-exposure
Ophiostoma piceae (387E) alone	4.33 a	5.00 a
O. piceae + *Gliocladium roseum* (321M)	2.67 b	1.33 b
O. piceae + Na_2CO_3	3.33 ab	5.00 a
O. piceae + *G. roseum* + Na_2CO_3	0.00 c	0.00 c
Aureobasidium pullulans (132S) alone	5.00 a	5.00 a
A. pullulans + *G. roseum*	4.00 b	1.00 c
A. pullulans + Na_2CO_3	3.33 b	3.33 b
A. pullulans + *G. roseum* + Na_2CO_3	0.00 c	0.00 d
Ceratocystis adiposa (251E) alone	4.33 a	5.00 a
C. adiposa + *G. roseum*	4.33 a	0.00 c
C. adiposa + Na_2CO_3	4.33 a	4.33 b
C. adiposa + *G. roseum* + Na_2CO_3	0.00 b	0.00 c
Alternaria alternata (2G) alone	4.00 b	4.00 a
A. alternata + *G. roseum*	2.00 c	0.67 b
A. alternata + Na_2CO_3	5.00 a	4.00 a
A. alternata + *G. roseum* + Na_2CO_3	1.67 c	0.33 b
Cladosporium sphaerospermum (806B) alone	5.00 a	5.00 a
C. sphaerospermum + *G. roseum*	4.00 b	3.67 b
C. sphaerospermum + Na_2CO_3	2.33 c	3.00 b
C. sphaerospermum + *G. roseum* + Na_2CO_3	0.00 d	0.00 c

[a] Data are means of three replicates. Means followed by the same letters in a treatment section are not significantly ($p = 0.05$) different from each other by Scheffe's test.

Table 4.—Growth of moulds and sapstaining fungi on sapwood wafers of sugar maple following different treatments.

Treatment	Treating type[a]	Average score[b]	Clean pieces	Acceptable pieces
			------ % -----	
Sapstaining fungi (SF) alone	SI	4.67 (0.52)	0	0
SF + *Gliocladium roseum* (321M)		1.17 (0.41)	0	100
SF + 5% Na_2CO_3		2.83 (0.98)	0	17
SF + *G. roseum* + 5% Na_2CO_3		0.00 (0.00)	100	100
Untreated control		3.50 (0.84)	0	17
Sapstaining fungi (SF) alone	PI	3.83 (0.41)	0	0
SF + *Gliocladium roseum* (321M)		0.17 (0.41)	83	100
SF + 5% Na_2CO_3		2.00 (1.26)	17	50
SF + *G. roseum* + 5% Na_2CO_3		0.00 (0.00)	100	100
Untreated control		3.50 (0.55)	0	0

[a] SI = Sapstaining fungal spore soup was applied onto wood wafers immediately following the treatment; PI = Sapstaining fungal spore soup was applied onto wood wafers 3 days later following the treatment.

[b] Values are means of six replicates with standard deviation in parentheses.

Table 5.—Growth of moulds and sapstaining fungi on sapwood wafers of white pine following different treatments.

Treatment	Treating type[a]	Average score[b]	Clean pieces	Acceptable pieces
			------ % -----	
Sapstaining fungi (SF) alone	SI	4.67 (0.52)	0	0
SF + *Gliocladium roseum* (321M)		3.67 (1.12)	0	17
SF + 5% Na_2CO_3		0.83 (0.98)	50	100
SF + *G. roseum* + 5% Na_2CO_3		0.00 (0.00)	100	100
Untreated control		4.17 (1.33)	0	17
Sapstaining fungi (SF) alone	PI	4.67 (0.52)	0	0
SF + *Gliocladium roseum* (321M)		1.67 (1.86)	50	50
SF + 5% Na_2CO_3		1.50 (1.64)	50	50
SF + *G. roseum* + 5% Na_2CO_3		0.17 (0.41)	83	100
Untreated control		5.00 (0.00)	0	0

[a] SI = Sapstaining fungal spore soup was applied onto wood wafers immediately following the treatment; PI = Sapstaining fungal spore soup was applied onto wood wafers 3 days later following the treatment.

[b] Values are means of six replicates with standard deviation in parentheses.

activities of *G. roseum* were strongly enhanced by the treatment of sodium carbonate solution on the medium. This demonstrates that an alkaline condition can help *G. roseum* establish itself well on the host and may stimulate this fungus producing more antagonistic compounds against other microorganisms.

Testing on Wood Wafers

On sugar maple wood, when sapstaining fungi were inoculated onto wafers immediately following the treatments with different control agents, two treatments, *G. roseum* alone and 5 percent sodium carbonate plus *G. roseum*, gave a satisfactory level of protection with 100 percent acceptable pieces during an 8-week testing period (Table 4). However, wood treated with the former resulted 0 percent clean pieces, while it was 100 percent clean by the treatment with the latter. Similar results were obtained from the test when sapstaining fungi were inoculated onto wood wafers 3 days later following the treatments with different control agents, but clean pieces of wood samples increased from 0 percent in the first set of the test to 83 percent in this test by the treatment of *G. roseum* alone (Table 4).

On white pine wood, when sapstaining fungi were inoculated onto wafers immediately following the treatments with different control agents, samples treated with 5 percent sodium carbonate or with 5 percent sodium carbonate plus *G. roseum* gave a satisfactory level of protection with 100 percent acceptable pieces (Table 5). The treatment of *G. roseum* alone only yielded 17 percent acceptable pieces. When sapstaining fungi were inoculated onto wafers 3 days later following the treatments with different control agents, only one treatment, 5 percent sodium carbonate plus *G. roseum*, gave satisfactory protection with 100 percent acceptable pieces (Table 5). The treatments with *G. roseum* alone or with 5 percent sodium carbonate alone only resulted 50 percent acceptable pieces.

Conclusions

The potential biocontrol agent, *G. roseum*, was more alkali-tolerant for both its spore germination and mycelial growth than sapstaining fungi. Simultaneous application of *G. roseum* and 5 percent sodium carbonate on malt extract agar totally eliminated growth from most sapstaining fungi tested. On sugar maple sapwood, though simultaneously applied *G. roseum* with 5 percent sodium carbonate further reduced the severity of infection and increased clean pieces from treated wood, the treatment of *G. roseum* alone provided satisfactory protection from stain during an 8-week testing period. On white pine sapwood, simultaneous application of *G. roseum* and 5 percent sodium carbonate significantly enhanced the degree of protection afforded by bioprotectant treatment alone. Further experiments will be extended to field conditions

Literature Cited

1. Bruce, A. and J. L. Highley. 1991. Control of growth of wood decay basidiomycetes by *Trichoderma* spp. and other potentially antagonistic fungi. Forest Prod. J. 42(2):63-67.
2. Dawson-Andoh, B.E. and J.J. Morrell. 1992. Enhancing the performance of bioprotection agents by pretreatment with waterborne salts. Wood and Fiber Sci. 24(3):347-352.
3. Dennis, J.J. 1985. Effect of pH and temperature on in vitro growth of ectomycorrhizal fungi. Information Report BC-X-273, Pacific Forestry Centre, Canada Forestry Service. 19 pp.
4. Freitag, M., J.J. Morrell, and A. Bruce. 1991. Biological protection of wood: status and prospects. Biodeterioration Abstr. 5(1):1-13.
5. Hung, L.L. and J.M. Trappe. 1983. Growth variation between and within species of ectomycorrhizal fungi in response to pH in vitro. Mycologia 75(2):234-241.
6. Laiho, O. 1970. *Paxillus involutus* as a mycorrhizal symbiont of forest trees. Acta Forest.Fenn. vol. 106. 72 pp.
7. McAfee, B. and M. Gignac. 1993. Two stage method for the protection of lumber against sapstain. U. S. Patent, Serial No. 08/281, 776.
8. Morrell, J. J. and C.M. Sexton. 1990. Evaluation of a biological agent for controlling basidiomycete attack of Douglas-fir and southern pine. Wood and Fiber Sci. 22(1):10-21.
9. Morrell, J.J. and C.M. Sexton. 1992. Effect of nutrient regimes, temperature, pH, and wood sterilization method on performance of selected bioprotectants against wood staining fungi. International Research Group on Wood Preservation Document No IRG/WP/1551. Harrogate, UK.
10. Morris, P.I., D.J. Dickinson, and J. F. Levy. 1984. The nature and control of decay in creosoted electricity poles. Rec. 1984 Annu. Conv. Br. Wood Preserv. Assoc. p. 42-53.
11. SAS Institute Inc. 1989. SAS/STAT user's guide, version 6 edition. SAS Institute Inc., Cary, N.C.
12. Seifert, K.A., C. Breuil, L. Rossignol, M. Best, and J.N. Saddler. 1988. Screening for microorganisms with the potential for biological control of sapstain on unseasoned lumber. Mat. und Org. 23(2):81-95.
13. Stranks, D.W. 1976. Scytalidin. hyalodendrin, cryptosporiopsin: antibiotics for prevention of blue stain in white pine sapwood. Wood Science 9(2):110-112.

Elimination of Sapstain in Radiata Pine Logs Exported from New Zealand

Tony Price

Carter Holt Harvey Ltd. and Fletcher Challenge Forests Ltd. have substantial radiata pine forest holdings in New Zealand and Chile. They are jointly funding several research projects to achieve this goal.

Approximately 20 percent of the annual harvest of logs is presently exported from New Zealand. This is a difficult task due to the long time delay from felling to processing and the environment prevailing on-board ships crossing the equator and in S.E. Asian log yards.

It is expected that an integrated control system will be necessary to achieve long-term protection. The following areas are being studied.

1. Improvements in present practice—harvesting, handling, and fungicide selection and application.
2. Use of natural products as a fungicide.
3. Moisture control.
4. Biological control.

The following figures describe the global radiata pine estate and New Zealand aspects. They are correct as of October 1996.

Global Radiata Pine Forest Estate

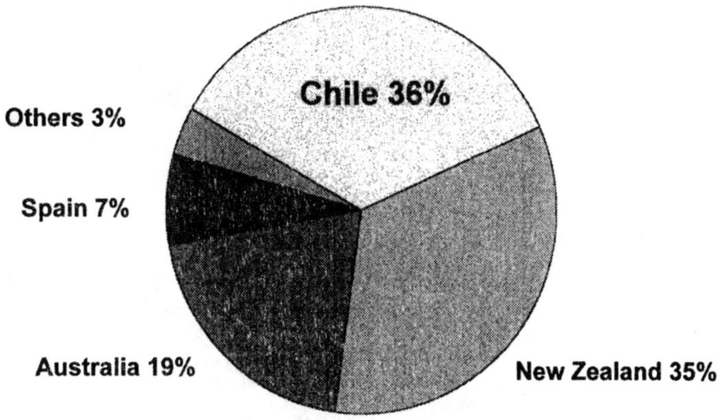

Estimated Total Area 3.8 million ha

Tony Price, PhD, Fletcher Challenge Forests, Ltd., Rotoura, New Zealand.

N Z Planted Production Forest Area by Species

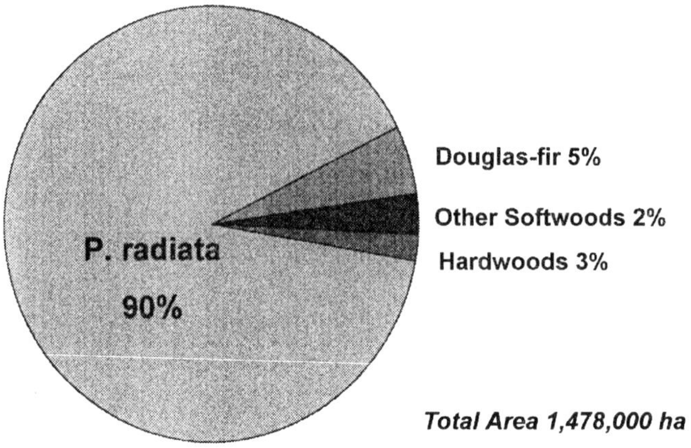

N Z Plantation Forest Ownership

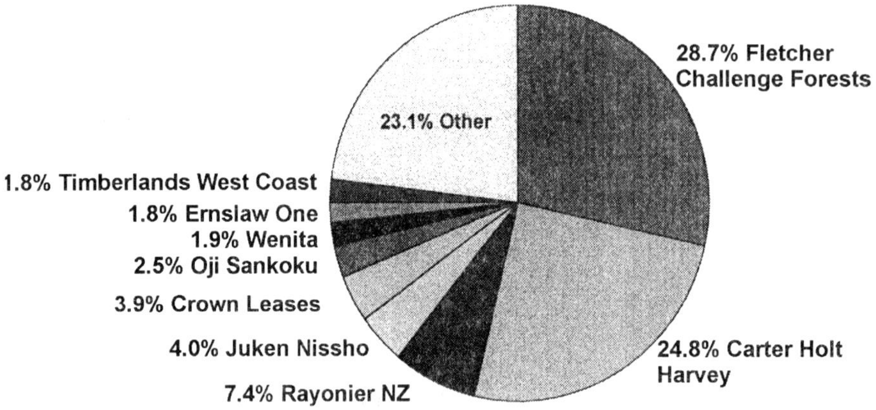

Total New Zealand Export Earnings

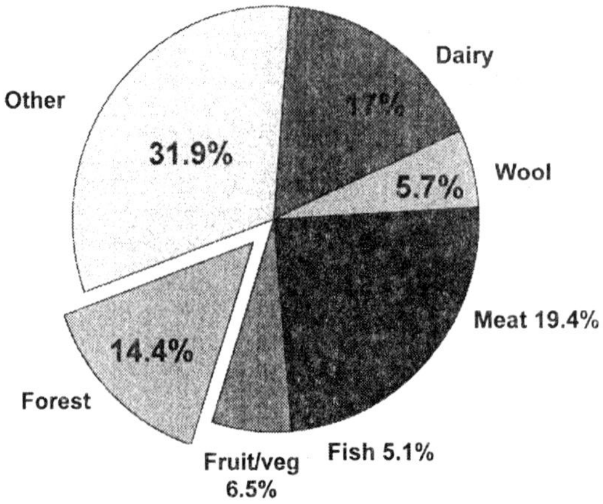

PLANTATION FORESTRY NORTH ISLAND

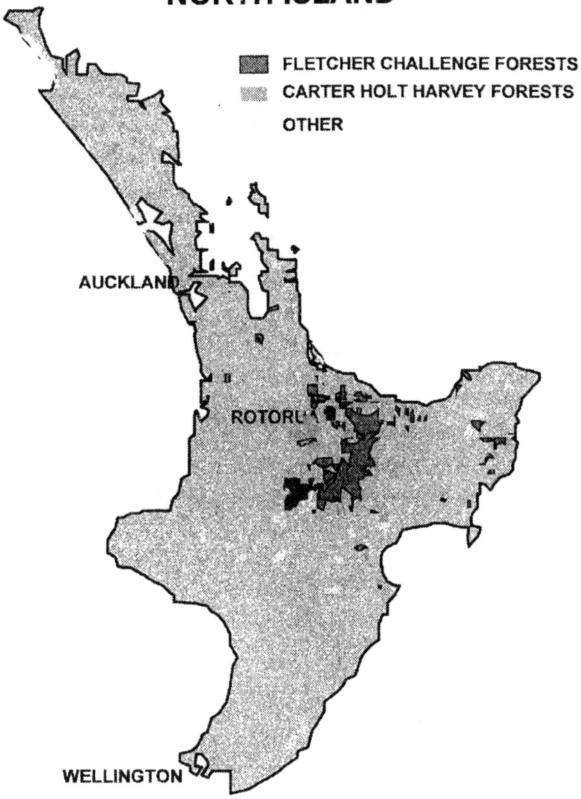

PLANTATION FORESTRY SOUTH ISLAND

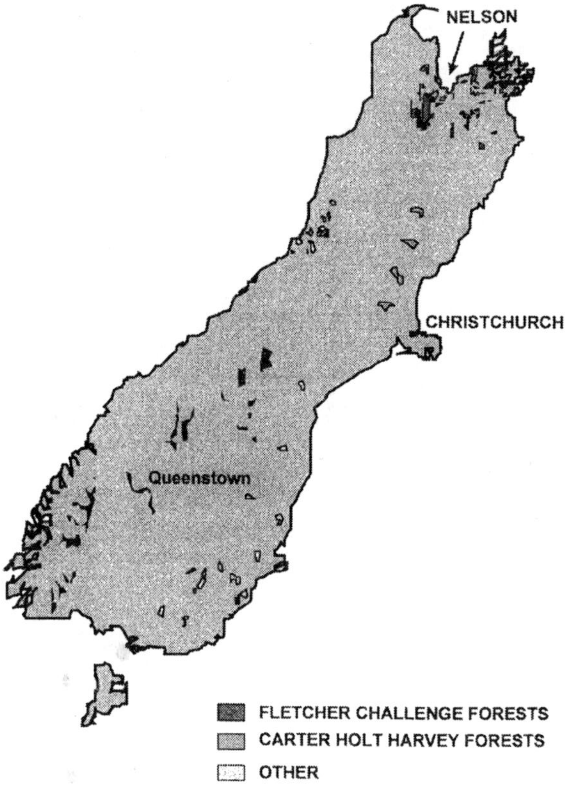

New Zealand Plantation Projected Annual Harvest (million m³)

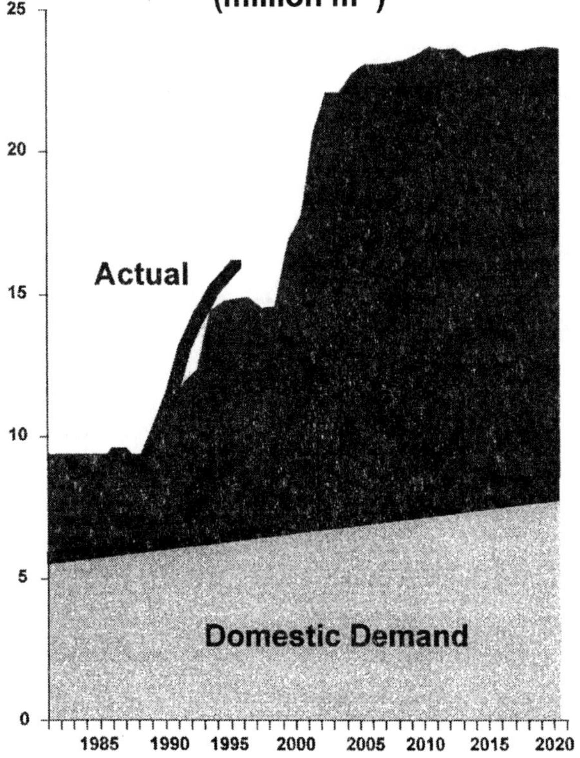

TYPICAL NZ LOG OUT-TURN Radiata Pine Age 30 from Direct Sawlog Regime

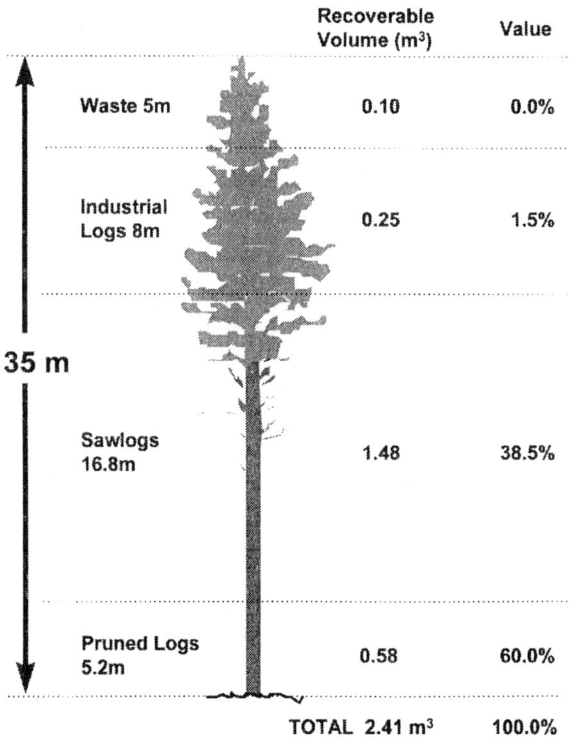

Survey of Sapstain Organisms in New Zealand and Albino Anti-Sapstain Fungi

Roberta L. Farrell, Esther Hadar, Stuart J. Kay, Robert A. Blanchette, and Thomas C. Harrington

Abstract

A 4-year study commenced in August 1996 with the objectives of surveying and identifying New Zealand sapstaining organisms, investigating the cause(s) of sapstain in radiata pine, and developing an albino fungal product to control sapstain.

The results of the first year have shown that *Sphaeropsis sapinea* (= *Diplodia pinea*) is found throughout the majority of the pine plantations in New Zealand and in about one third of the native forests sites sampled. *S. sapinea* exists extensively as an endophyte on radiata pine at pruned sites and under the bark of healthy trees from age 1 to maturity, about 25 years. *S. sapinea* has also been found on material from the forest floor including branches, cones, leaf litter, and wood. Members of the family Ophiostomataceae were found in about one quarter of pine plantations, in about one third of the non-pine sites sampled, and in virtually all samples from pine processing sites. *Ophiostoma piceae* C and H (the conifer and hardwood forms) were both isolated from radiata pine, which makes the New Zealand situation atypical from the European situation. *O. ips*, *O. piliferum*, *O. pluriannulatum*, *O. stenocerus*, *Leptographium procerum*, and *L. truncatum* as well as isolates of the anamorph genus *Graphium*, one distinguished as having black veined synnema and the other with red-brown synnema, have also been identified. The sexual stages of the *Graphium* species with red-brown synnema, but not the black veined *Graphium* species, have been produced in culture. Initial studies on the biocontrol of *S. sapinea*, with light *Ophiostoma* sp. and *Graphium* sp. isolates, has shown good anti-sapstain protection of radiata pine. The eventual goal of the program is the development of an albino Ophiostomataceae organism used as an anti-sapstain product for radiata pine.

Introduction

Radiata pine (*Pinus radiata*) sapwood is highly susceptible to fungal attack especially by sapstain fungi. Currently an efficacious, economic, and environmentally sound method of sapstain prevention is being sought by the New Zealand Forest Industry. The major markets for export logs from New Zealand radiata pine include Japan, Korea, the United States, and the Philippines, and with warm moist conditions and lengthy transport times, prevention of sapstain is crucially important. To control sapstain, an understanding of the causative agent(s) of sapstain, timing of appearance, and aggravating factors promoting stain is necessary.

Studies of sapstain organisms in New Zealand were conducted by Butcher and his colleagues in the late 1960s through 1980s and then in 1982 by J. Reid (6,10,11). These studies sampled only a few selected areas, mainly on the North Island. Wood samples were collected from both native and introduced tree species and studied for wood staining fungi (10,11). Culture descriptions were based on colony morphologies of isolates grown in a controlled environment. The following sapstain organisms were identified: *Sphaeronaemella fimicola*, *Ceratocystiopsis falcata*, *Ophiostoma piceaperdum*, *O. ips*, *O. novo-*

Roberta L. Farrell, Department of Biological Sciences, University of Waikato, Hamilton, New Zealand, Esther Hadar, Extension Service, Ministry of Agriculture, Rehovet, Israel, Yitzhak Hadar, Faculty of Agriculture, The Hebrew University of Jerusalem, Rehovet, Israel, Stuart J. Kay, Natural Products Group, The Horticultural and Food Research Institute of New Zealand, Ruakura Research Centre, Hamilton, New Zealand, Robert A. Blanchette, Department of Plant Pathology, University of Minnesota, St. Paul, Minnesota, and Thomas C. Harrington, Department of Plant Pathology, Iowa State University, Ames, Iowa.

zealandiae, *C. piceae*, *C. coronata*, *C. rostrocoronata*, and *C. piliferum*. This study did not link these organisms with their overall New Zealand distribution nor their contribution to sapstain in radiata pine.

Trees growing under stress or with pruning wounds have been shown to develop of *S. sapinea* infections (7). Sapstain caused by *S. sapinea* develops under bark, but more particularly on wood exposed by bark removal during handling. Severe internal sapstain caused by pigmented fungal hyphae penetrating wood rays develops within 3 months (8). In studies with roundwood storage, *S. sapinea* 'was by far the most important cause of sapstain', presumably carried over from infected logs and representing 70 percent of all isolations of fungi from sapstained wood, confirming the earlier reports of Birch (3). For roundwood, minor species were *Ceratocystis* spp. (regarded as of major importance by Yeates, 13)) and *Alternaria alternata* (6). *Ceratocystis* spp. were shown to be the main fungi causing sapstain through new infections of freshly sawn timber.

Radiata pine is of paramount importance in New Zealand accounting for approximately 95 percent of all exotic forestry plantings. The forestry and forest product industries produce about 5.7 percent of New Zealand's GDP and are the third most important export earner. The lack of a 'robust', cost effective, and environmentally acceptable preventative of sapstain and the costs associated, has led to the search for new anti-sapstain products, including biocontrol agents.

Technology developed by Sandoz Chemicals Biotech Research Corporation (now Biotech Division of Clariant Corporation), initiated in 1987, showed that stable albino isolates could be developed by classical genetic methods from wild-type *Ophiostoma piliferum* strains isolated from southern yellow pine (*Pinus taeda*) collected primarily in South Carolina and Virginia. The product from this technology, Cartapip®, was initially developed for pitch (i.e., triglycerides, fatty and resin acids, sterols and waxes) reduction and applied to chips as a fungal inoculum prior to thermomechanical pulping. An added advantage was the improved brightness of the chips due to the colonization of the albino Cartapip product and biocontrol i.e., prevention of colonization by sapstaining organisms (9). One strain, Cartapip 97, obtained by single ascospore isolations, was shown to be blocked in the 1,8-dihydroxynaphthalene (DHN) melanin synthetic pathway such that the intermediate scytalone could not be made (14). Later work showed that *O. piceae* albino strains could also be generated by the same methodology (12). Biocontrol field studies demonstrated the effectiveness of the albino *O. piliferum* protecting cut red pine (*Pinus resinosa*) logs from blue stain (1,2).

The present study has been undertaken with three objectives:

1. Broadly survey sapstain organisms in New Zealand with variables including geographic and temporal distribution during four seasons, tree age and health, harvesting and storage impact.
2. Investigate the causes of sapstain in radiata pine logs.
3. Develop, from the Ophiostomataceae isolates of New Zealand, an albino anti-sapstain product for radiata pine logs.

This paper describes results from the first 9 months of the survey in New Zealand and initial positive biocontrol results against *S. sapinea* with New Zealand Ophiostomataceae isolates.

Material and Methods

Survey

Sampling has been done every 3 months from about 100 sites distributed throughout North and South Islands of New Zealand. The survey extends from Southern Hemisphere Winter 1996 to Winter 1998; thus a total of about 800 site samplings will be taken. Samples are first placed in plastic bags, sealed, and used for culture isolations. From each site the following is sampled:

1. from the forest floor: branches, twigs, needles, cones and wood
2. from standing trees: bark, green needles, and sapwood, when possible bark and sapwood from pruned or wounded sites is also taken
3. from recently felled trees in the forest, at mill and ports sites in New Zealand, or from export logs from customers in the Philippines, Korea, Japan, or the United States a large biscuit section, at least 30 centimeters in length, of sapwood and bark is taken.

Radiata pine plantations, from first to fourth rotation, as well as native forest, mill and port sites, urban areas, and horticultural and agricultural land have all been sampled.

Culture Isolations

For fungal isolations, bark was carefully separated from the underlying sapwood surface, and slivers from the inner face of the bark and the wood surface were isolated aseptically using standard techniques. Sapwood was sampled from the exterior surface, and then was soaked in 2 percent hypochlorite solution for 1 minute, rinsed in sterile water, and with a sterile scalpel slivers were taken from the interior. Sample pieces were placed onto two different plates both consisting of YMA (0.2% yeast extract, 2.0% malt extract, and 2.0% agar), supplemented with 200 micrograms per ml chloramphenicol and 100 micrograms per ml streptomycin sulphate and one set also supplemented with 400 micrograms per ml cycloheximide.

Inoculated plates were incubated in a darkened growth chamber at 25°C for 4 to 21 days. Any resulting cultures were aseptically transferred onto fresh plates with and without cycloheximide. All cultures suspected of being *S. sapinea* were inoculated onto sterile radiata pine needles and maintained under ultraviolet light at ambient temperature for 2 weeks, after which they were examined microscopically in sterile water for the presence of pycnidia and spore release.

Cultures of Ophiostomataceae were identified on the basis of morphological features into putative species. Molecular characterizations, particularly DNA sequences of the internal transcribed spacer regions of ribosomal DNA, and mating compatibilities with tester strains have been used to confirm species identifications.

In order to determine the color and the extent of the stain produced by an isolate, mycelium and spores from cultures grown in YM medium overnight in shake flasks were inoculated onto gamma irradiated sterile radiata pine cubes (2 by 2 by 2 cm). The inoculated cubes were incubated in the dark at 5°, 13°, and 25°C for 3 weeks.

Morphological structures important in the identification of isolates were stained with trypan blue and photographed using phase-contrast microscopy.

Cultures were maintained at the Department of Biological Sciences, University of Waikato, at –70°C in 20 percent glycerol or, after growth, on sterile radiata pine cubes.

Competition Experiments

Competition experiments were carried out on gamma irradiated sterile radiata pine cubes (2 by 2 by 2 cm) at 100 percent relative humidity. Cultures, produced by growing overnight in YM in shake culture, were inoculated onto the cubes at a dose of approximately 10^6 colony forming units (cfu) per cube. Ophiostomataceae isolates were inoculated and incubated at 25°C for 3 or 7 days prior to *S. sapinea* inoculation. Inoculated cubes were maintained in a closed sterile container for 5 weeks at 25°C. Assessment of competition was carried out at 3 and 5 weeks using a stain value scale (intensity of stain) and percentage coverage by stain. The assessment examined the internal discoloration of the cubes and was repeated three times by separate assessors.

A score of 0 to 5 for the stain value indicated the following:

0 = fresh, completely unstained, clean wood
1 = palest grey stain
2 = pale grey stain
3 = grey stain
4 = dark grey to black stain
5 = dark black

Percentage coverage was expressed in multiples of 10 from 0 to 100 percent of the area showing stain.

Pitch Assessment

Pitch was determined by inoculation of the isolates, at a dose of 10^7 cfu per kilogram ovendry weight, onto sterile gamma irradiated radiata pine chips obtained courtesy of the Waipa Saw Mill, Rotorua, New Zealand. After 3 weeks of incubation at 25°C, extraction with dichloromethane was done in a Soxtec extractor by standard procedures by the Forest Research Institute Analytical Chemistry Laboratory.

Results and Discussion

In the 3 months of Austral Spring (September, October, November) and 3 months of Austral Summer (December, January, February) about 200 sites were sampled. About 75 percent of the sampling was carried out at radiata pine sites and the rest at non-pine sites. As mentioned the pine sites included single trees to large plantations from age classes of 1 to 26 years, shelter belts, wind fallen trees, and radiata pine processing sites identified as mills or ports. Variables in the sampling for the survey included temporal (four seasons), geographic location, harvesting and transport conditions, weather, soil, vegetation, pruned versus wounded trees, and health of trees.

The dominant sapstain fungi have been identified as *Sphaeropsis sapinea* and members of the Ophiostomataceae. Mould genera isolated included *Alternaria, Fusarium, Rhizopus, Penicillium, Trichoderma,* and *Verticillium. Geotrichum* has also been found throughout North and South Islands. Table 1 shows the distribution of these fungi per radiata pine plantations, processing

Table 1.—Isolations from New Zealand summer survey. Percentage of total number sites sampled positive for organism.

	Pine plantation sites	Pine processing sites	Non-pine sites
Sphaeropsis	88	60	38
Ophiostoma	28	100	31
Alternaria	69	25	42
Geotrichum	1	0	4
Verticillium	37	40	35

Other common fungi: *Trichoderma, Penicillium, Rhizopus,* and *Fusarium.*

sites, and at non-pine sites during the summer months. *S. sapinea* was present in 88 percent of all the pine plantation sites, a positive site can be with *S. sapinea* only found on needles or cones and not necessarily on wood. *S. sapinea* was absent from nine sites sampled on South Island. Ophiostomataceae isolates were identified in 28 percent of pine plantation sites sampled.

At radiata pine processing sites, the percentage of sites with *S. sapinea* decreased to 60 percent, whereas 100 percent of sites sampled showed presence of Ophiostomataceae. The number of positive sites for *S. sapinea* and Ophiostomataceae isolates from non-pine sites was similar, with about one third of the sites positive.

A similar trend has been observed in mature trees felled in three forests distributed on North Island, the forests were Kinleith Forest, Mahurangi Forest, and Whitford Forest. At felling, *S. sapinea* was isolated from the bark next to the cambium, but not in the sapwood (except in the case of a spar, standing dead tree, which had extensive sapstained wood upon felling and from which *S. sapinea* was isolated). The logs were progressively sampled over time, *S. sapinea* was isolated from only the bark until about week 2 at which time it was also isolated from sapwood, at week 3 *S. sapinea* was isolated from the bark next to the cambium, and *S. sapinea* and Ophiostomataceae isolates were both isolated from inside sapwood. The presence of *S. sapinea* did not stop the colonization of sapwood by Ophiostomataceae.

The spores of New Zealand isolates of *S. sapinea*, observed by phase microscopy, resembled the Type A Northern Hemisphere *S. sapinea* spores. Type A has been shown to be the more aggressive form of *S. sapinea* in the Northern Hemisphere (4).

Initial sampling at ports has shown that 36 percent of logs sampled were positive for *S. sapinea* and 64 percent positive for Ophiostomataceae. Approximately a quarter of the logs sampled were positive for both *S. sapinea* and Ophiostomataceae isolates.

A list of the Ophiostomataceae species identified in the first 9 months of the survey are given in Table 2.

Both mating types of *O. piceae* C (OPC) and *O. piceae* H (OPH) have been identified in the survey and are widely geographically distributed. OPC and OPH have both been isolated from radiata pine and in some cases from the same piece of radiata pine wood. Their co-occurrence on pine differs from reports from the Northern Hemisphere, where OPC is more common on conifers and OPH on hardwoods (5). In New Zealand OPH has also been isolated from *Cupressocyparis macrocarpa*.

Isolates of the anamorph genus *Graphium* with black veins at the top of the synnema have thus far only been isolated in New Zealand. These isolates are not sexually compatible among themselves nor with tester strains of any other *Ophiostoma* species with *Graphium* anamorphs. Based on the ITS sequences of the rDNA region, this species is close to OPC (T. Harrington, pers. comm.). Similarly, isolates of the *Graphium* species with red-brown synnema have an ITS sequence which is close to that of OPC; however, in contrast it is also identical to isolates from the UK and the United States, and the New Zealand strains show sexual compatibility with UK strains. The perithecia and ascospores are similar to, but distinct from, *O. piceae*.

In order to develop an albino biocontrol anti-sapstain product for both *S. sapinea* and the wild-type, dark-staining Ophiostomataceae, their growth in radiata pine has been studied. Growth of *S. sapinea* occurred throughout ray parenchyma cells and resin ducts of radiata pine. Colonization patterns in the wood appeared similar for *S. sapinea* and Ophiostomataceae isolates studied by scanning electron microscopy.

Table 2.—Ophiostomatacaea sampled and identified.[a]

Graphium sp. with black veined synnema[b]
Graphium sp. with red-brown synnema[b]
Leptographium truncatum
Leptographium procerum
Ophiostoma ips
Ophiostoma piceae conifer form (OPC)
Ophiostoma piceae hardwood form (OPH) = *Ophiostoma querci*
Ophiostoma piliferum
Ophiostoma pluriannulatum
Ophiostoma stenocerus

[a] Samples identified from isolates made in New Zealand from September 1996 to March 1997.

[b] Unidentified species, closely related to *Ophiostoma piceae* conifer form.

Table 3.—Pitch reduction in radiata pine by New Zealand isolates.

Isolate	DCM extractives	DCM decrease vs control
	(g/100 g wood)	(%)
Control	0.75	
Control	0.72	
O. piceae C Isolate 40	0.50	32
Graphium 'black vein synnema' Isolate 67	0.60	19
O. piceae H Isolate 19	0.61	18
O. piceae H Isolate 27	0.63	15
S. sapinea Isolate 35	0.53	28
S. sapinea Isolate 4	0.59	20

Table 4.—Competition experiment on cubes.

	Intensity and amount of discoloration of the internal tissue of radiata pine cubes			
	Week 3		Week 5	
Sample	Mean stain value	% stain coverage	Mean stain value	% stain coverage
Control	0.0	0	0.0	0
S. sapinea isolate 4 alone	3.7	55	4.0	95
Isolate 67 alone	0.7	11	0.5	30
Isolate 67 7 days prior to S. *sapinea* inoculation	1.3	78	–	–
Isolate 67 3 days prior to *S. sapinea* inoculation	0.9	71	1.0	82
Isolate 68 alone	0.2	33	0.0	0
Isolate 68 7 days prior to S. *sapinea* inoculation	0.6	64	0.75	86
Isolate 68 3 days prior to *S. sapinea* inoculation	0.7	70	0.75	95
Isolate 19 alone	0.0	3	0.0	0
Isolate 19 7 days prior to S. *sapinea* inoculation	0.0	0	0.4	56
Isolate 19 3 days prior to *S. sapinea* inoculation	0.8	63	0.75	40
Isolate 40 alone	0.8	78	0.9	89
Isolate 40 7 days prior to S. *sapinea* inoculation	0.9	58	0.5	60
Isolate 40 3 days prior to *S. sapinea* inoculation	0.9	73	1.4	100

Since both Ophiostomataceae and *S. sapinea* isolates were observed in resin canals, and with the understanding of the Cartapip product's ability to decrease pitch, New Zealand isolates were tested for their ability to reduce pitch after inoculation onto previously sterilized gamma irradiated radiata pine wood chips. Six isolates, two *S. sapinea* and four Ophiostomataceae, were chosen for a pitch reduction experiment. Of the two *S. sapinea* isolates, S. sapinea isolate 4 was isolated from forest litter on the ground in Northland and *S. sapinea* isolate 35 was isolated from bark in a live tree in Kinleith Forest. Two controls were maintained with no fungi inoculated onto the chips. The results are given in Table 3.

Isolate OPC isolate 40 and *S. sapinea* isolate 35 gave significant pitch reduction. The other Ophiostomataceae strains gave about the same range of low DCM reduction. The pitch decrease by the endophytic strain supported the observations of *S. sapinea* hyphae in the radiata pine resin canal and consequently its involvement in pitch reduction.

The causes of sapstain in New Zealand radiata pine, especially concerning the almost ubiquitous in the forest *S. pineae*, and Ophiostomataceae, an aggressive invader after trees are felled and at processing sites, will continue.

Development of a successful albino product requires an organism with no stain/color to out compete and stop the appearance of staining organisms on wood.

An experiment of gamma radiated radiata pine cubes was conducted with four different New Zealand Ophiostomataceae isolates against *S. sapinea*. The Ophiostomataceae isolates chosen for this experiment were *Graphium* black veined synnema isolate 67, *Graphium* red-brown synnema isolate 68, OPH isolate 19, and OPC isolate 40. All produce minimal stain. Results of the competition experiment are given in Table 4.

The competition experiment found that the biocontrol of *S. sapinea* was achieved with all of the Ophiostomataceae isolates tested. *S. sapinea* alone produced significant stain. Competition experiments against *S. sapinea* with native pale strains will be continued in order to determine the potential of the various Ophiostomataceae to compete against New Zealand *S. sapinea* and New Zealand black Ophiostomataceae strains. The best strains that stop *S. sapinea* are the candidates for mating and creating the albino anti-sapstain product.

Acknowledgments

The authors acknowledge the generous support of Fletcher Challenge Forests and Carter Holt Harvey Forests. The authors also acknowledge Clariant Corporation for its collaboration with this work. Excellent technical assistance has been provided by Shona Duncan, Arvina Ram, Doug McNew, and Barry O'Brian for which the authors are very grateful.

Literature Cited

1. Behrendt, C.J., R.A. Blanchette, and R.L. Farrell. 1995a. Biological Control of Blue-Stain Fungi in Wood. Phytopath. 85:92-97.
2. Behrendt, C.J., R.A. Blanchette, and R.L. Farrell. 1995b. An integrated approach, using biological and chemical control, to prevent blue stain in pine logs. Can J. Bot. 73:613-619.

3. Birch, T.T.C. 1936. Diplodia pinea in New Zealand. Bulletin New Zealand Forest Service, No. 8, 32 p.
4. Blodgett, J.T. and G.R. Stanosz. 1997. *Sphaeropsis sapinea* morphotypes differ in aggresiveness, but both infect non-wounded red or jack pines. Plant Disease pp. 143-147.
5. Brasier, C.M. and S.A. Kirk. 1993. Sibling Species within *Ophiostoma piceae*. Mycol. Res. 97:811-816.
6. Butcher, J.A. and J.A. Drysdale. 1991. Biodeterioration and Natural Durability. In: Properties and Uses of New Zealand Radiata Pine J.A. Kininmonth and L.J. Whitehouse (eds.).
7. Chou, C.K.S. 1984. Diplodia leader dieback, Diplodia crown wilt, Diplodia whorl canker. Forest Pathology in New Zealand No. 7. 2 p.
8. Drysdale, J.A., M.E. Hedley, and J.A. Butcher. 1986. An evaluation of chemical treatments for the protection of radiata pine logs from fungal degrade. The International Research group on Wood Preservation Document No. WP/IRG/3377.
9. Farrell, R.L., R.A. Blanchette, T.S. Brush, Y. Hadar, S. Iverson, K. Krisa, P.A. Wendler, and W. Zimmerman. 1993. Cartapip®: A biopulping product for control of pitch and resin acid problems in pulp mills. J. Biotechnol. 30:115-122.
10. Hutchison, L.J. and J. Reid. 1988a. Taxonomy of some potential wood-staining fungi from New Zealand 1. Ophiostomataceae. New Zealand Journal of Botany 26:63-81.
11. Hutchison, L.J. and J. Reid. 1988b. Taxonomy of some potential wood- staining fungi from New Zealand 2. Pyrenomycetes, Coelomycetes and Hyphomycetes. New Zealand Journal of Botany 26:83-98.
12. White, W. 1996. Master of Science Thesis.
13. Yeates, J.S. 1924. Sapstain in timber of *P. radiata* (insignis). New Zealand Forest Service circular No. 18.
14. Zimmerman, W.C., R.A. Blanchette, T.A. Burnes, and R.L. Farrell. 1993. Melanin and perithecial development in *Ophiostoma piliferum*, Mycologia. 87:857-863.

Chemical Control of Biological Stain: Past, Present, and Future

A. Byrne

Abstract

This paper reviews past and present industrial practices in controlling biological stain (sapstain, moulds, and decay) in sawn lumber. Issues and lessons learned are distilled and projections about future practices are made. Although an increased proportion of lumber is now kiln dried the need for chemical protection of export lumber will remain for the immediate future. The need for protection of logs from stain will probably increase. Protection of wood against stain requires that the problem be considered as a whole: the causal fungi, the control product, its application, and its monitoring on the wood surface. Understanding the fungi which cause staining problems has been rudimentary in the past but is currently improving. Increased knowledge in this area will be key to innovative control strategies. There has been an evolution of chemicals used to treat wood from broad spectrum biocides to mixtures of more benign chemicals. The use of biologicals to replace or partially replace chemicals is a possibility in the future. While application methods have improved since dipping was the standard procedure, hydraulic spraying could be made more effective by adopting a more highly technological approach to the application problem. Monitoring the application of control products is an important check to ensure that they are being applied evenly at a level required to control the pest organisms.

Introduction

Biological stain refers to the discolorations caused by sapstain fungi, moulds, and wood-decaying fungi which infect a tree after it is cut down and continue to develop while the wood cut from the tree is still moist. These organisms can deeply discolor the wood or make the wood surface unsightly with fungal growth.

This paper will draw from Canadian, mainly western Canadian, experience with chemical control of stain. The Canadian experience is relevant because Canada is the largest user of sapstain control products in the world (8). Additionally Canada has been a leader in tackling environmental issues and in developing application technology and monitoring techniques.

Unless it is controlled stain can cause large financial losses. Although the problems start to occur in logs, the focus in Canada has been on stain prevention in sawn lumber in which the problems become manifest. Lumber is and has been the largest requirement for stain control technology. The need for stain control in lumber derives from the fact that wood is stored at the sawmill, for example prior to kiln drying, or it is shipped, in green condition. Worldwide, although a higher proportion of lumber is now dried prior to shipping there is still a need for a substantial amount of lumber and also logs to be chemically treated. In the province of British Columbia alone about 5 million m^3 lumber with an estimated sales value of $2 billion was chemically treated with sapstain control products in 1996 (estimate based on Statistics Canada data). In addition an unknown amount of this lumber was treated twice and an unknown amount of lumber was treated prior to kiln drying.

Although this paper concentrates on lumber, the need to protect logs from stain is increasing around the world, particularly for the pines which are very susceptible.

There are three possible methods of protecting wood from stain: 1) physical methods – drying the wood to the point where fungi will not grow; 2) treatment with chemical pesticides; or 3) biological manipulation of the substrate, such as by using a biological pesticide. A fourth option would be to use a combination of physical, chemical, or biological methods.

A. Byrne, Wood Protection Scientist, Forintek Canada Corp., Vancouver, B.C., Canada.

For an industry which wishes to sell a green lumber product kiln drying is not regarded as an option. Unfortunately biological methods of controlling stain, while they are being investigated, are not yet practical. The industry therefore uses chemical pesticides as the only reliable method to control the problem on green wood.

As with controlling other pests the following 4-step sequence logically needs to be followed:

1. Understanding the pest organisms to be controlled;
2. Using the best control methods or products available;
3. Applying these control products properly;
4. Monitoring the effectiveness of the treatment through a quality control program.

The objective of this paper is to discuss each of these steps in the context of past and present practices. The paper will also address the main issues and draw lessons for future control strategies.

Understanding the Organisms: Past, Present, and Future

Knowledge of the organisms causing stain problems is far from complete. Often, the industry or research community does not know which are the main problem fungi and exactly when and how they infect. In the past, from the point of view of an industry with access to relatively potent chemicals, there was little apparent need to understand the causal organisms. A chemical was applied to the wood and if fungus did not grow the treatment was simply regarded as successful; if fungus grew the treatment was unsuccessful.

In the future we will need to know more about the problem fungi before we can make significant improvements in control strategies. There is a lot of excellent work presently being done in this field and enormous steps are being made in understanding both the staining organisms and their vectors. Some of the subjects being researched at present are taxonomy and genetics, aspects of fungal biology and ecology such as how they infect, how they colonize, and how they stain wood. The papers presented at this meeting are a good foundation for developing future, targeted control methods.

Using the Best Control Products

Although chemicals dominated in the past and still do at the present it is possible that biological products will be offered to control stain in the future. For this reason I refer to products rather than chemicals.

Past

Between the early 1900s and the 1980s stain control issues were reasonably clear cut. Protection requirements varied from weeks, for those air drying stock, to about a year for those exporting long distances. The control product needed to be cheap and easy to use by dissolving it in water. Issues that started to arise in the 1970s and 1980s were that workers were being exposed to chemicals and that the chemicals were escaping into the environment.

The control products used in the past started with salts such as sodium bicarbonate, sodium carbonate, and sodium borate (6). These salts were not fully effective especially for lumber which was exported or when close piling or block stacking was introduced.

In the 1930s potent biocides such as chlorophenols and organo-mercurials were adopted and because of their broad spectrum of toxicity they worked well. In Canada mercurials were abandoned in the early 1960s but chlorophenols continued on until the mid-1980s. The mood of the day was well captured by the trade names of some of the products offered, for example, Denicide, Diatox, Kilstain, and Permatox.

Three important lessons learned from the past are:

1. We need to use the most benign active ingredients feasible. Dioxins and chemicals which persist in the environment came under particular attack;
2. though various active ingredients were tested for many years in small scale trials (5) we learned the lesson that the active is only part of the control product and the formulation of the active is critical;
3. Worker (and other non-target organism) exposure must be minimized. This was done by adapting better handling and application practices and by providing workers with personal protective equipment.

Present

Canadian requirements of stain control products at the present are that they must be effective for several months rather than for a full year. They must also be safe for workers, as environmentally benign as feasible, and acceptable to customers.[1] These requirements reflect issues and lessons learned from the past.

A current issue is the general public's dislike and distrust of "chemicals." For example in the public press the word "toxic" often automatically appears before the word chemical. A further issue which has arisen in British

---

[1] For example in the mid-1980s German customers started to refuse to accept lumber treated with cholorphenates and B.C. pulp mills rejected chips which were contaminated with dioxins.

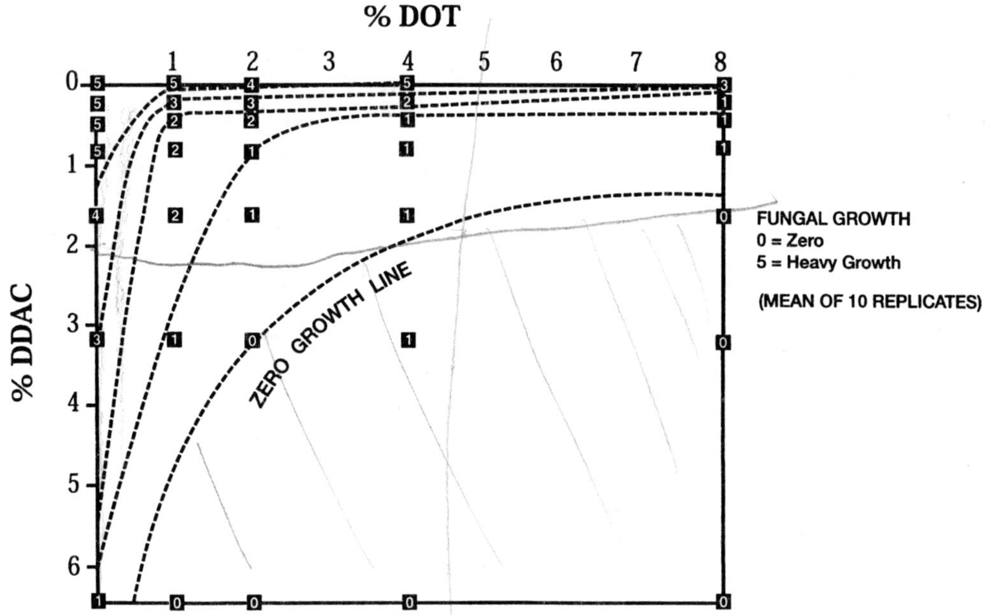

Figure 1.—Growth of fungi on combinations of disodium octaborate tetrahydrate (DOT) and didecyldimethyl-ammonium chloride (DDAC) — ASTM D4445 laboratory test.

Columbia is the need to involve stakeholders in decisions about what sapstain control products the industry can use and how they use them. Stakeholders include labor representatives, environmental advocates, and government regulators as well as chemical suppliers and the customers or users of their products. This type of stakeholder discussion could well start to occur in other parts of the world.

The chemical products currently being used worldwide are based on biocides with a narrower spectrum of activity than the earlier products. Mixtures of two active ingredients are starting to predominate and most of these active ingredients require relatively sophisticated formulation technology. In contrast with the names from the past we find the products are more likely to contain acronyms and simple numbers or to include the word "Brite" rather than "kil", "tox", "perm", and "cide." In Canada formulations called NP-1, F2, and QC3 are the most commonly used products, in order of current market share.

Compared with the older products five differences are notable:

1. Most of the modern ones are relatively expensive;
2. They are not as flexible with respect to fungal and wood species and protection times;
3. There are sometimes technical difficulties in their application;
4. The track record of some of the products is not well established;
5. Mixtures of active ingredients are common.

Regarding the last point, there are advantages of mixing active ingredients in formulations. If complementary active ingredients are chosen well and optimized there is potential that the efficacy of the whole is greater than the sum of the parts. This would not preclude parts of this mixture being biological in origin. Mixtures therefore enable us to use more benign actives. They also have the advantages that they can be tailored for particular circumstances and could (in theory) be flexibly altered to suit changing circumstances. Flexibility is more difficult in practice because of pesticide registration regulations.

Figure 1 summarizes a demonstration of synergism between two active ingredients in a laboratory experiment. The objective of this test was to optimize (i.e., maximize efficacy/$) a mixture of two biocides selected because of their acceptability for wood protection but also knowing that each had weaknesses against parts of the spectrum of fungi causing problems. The biocides are disodium octaborate tetrahydrate (DOT) and didecyldimethylammonium chloride (DDAC).

The ASTM D 4445 (1) standard screening test was used. This involves dipping small samples in the test solution and inoculating them with a spore suspension. In this case a mixture of fungi was used. After 4 weeks incubation fungal growth was rated on an integer scale of 0 to 5 (0 = very heavy fungal growth). In Figure 1 the average ratings are recorded on a matrix representing particular mixtures of DOT and DDAC increased in geometrical progression; lines linking points of equal growth have then been superimposed. These lines of approximately equal control bend convexly toward the origin illustrating that the combinations were more effective than the individual components. A mixture of approxi-

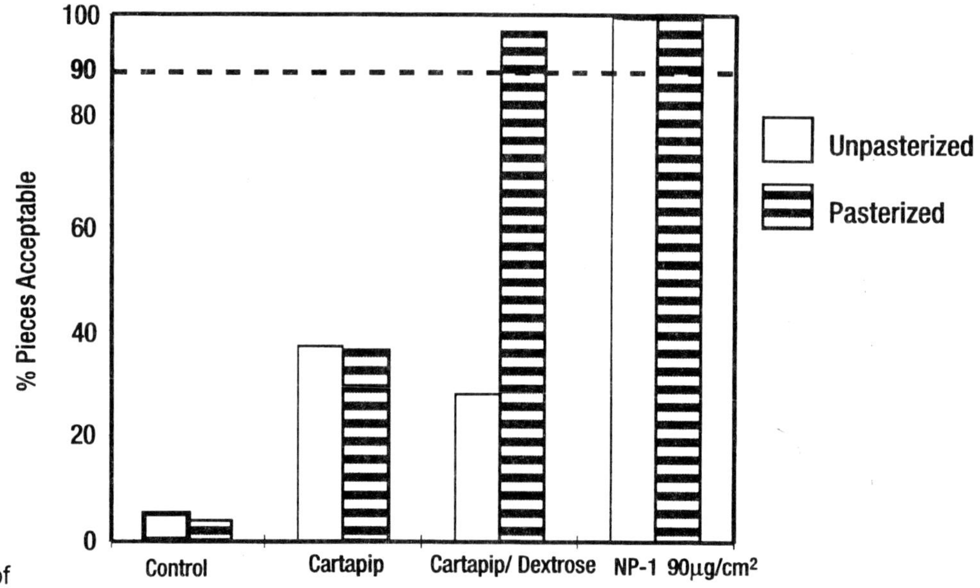

Figure 2.—1995 field test of Cartapip on pasteurized and unpasteurized wood.

mately 1:1 of the two actives gave the most protection for the least total amount of chemical. Adding lines representing equal cost of the two actives (not shown here) enables determination of the relative proportions of active which give the most protection for the lowest cost.

Future

The requirements of the ideal product of the future is a combination of all the most stringent of requirements to date. It must be:

1. Consistently effective over time spans of weeks to months on logs or lumber;
2. Cost effective;
3. Flexible in that the formulation may have to be compatibile with colorants, water repellents, and additives for non-microbial stain;
4. As environmentally benign as possible: not a threat to non-target life forms such as aquatic and soil organisms.

What might these products be? No doubt we will see new molecules, for example triazoles which have been developed by several chemical companies for the agricultural sector in recent years. In theory new molecules could be particularly targeted to vulnerable points in the life cycle of the fungi causing stain. However it is not very likely that new compounds will be developed specifically for protection of wood against stain.

The benefits of mixing actives were covered earlier. No doubt more complex mixtures and relatively sophisticated formulations of these will be developed. Some traditional fungicides, which have been overlooked because of their narrow spectrum of activity, could be used in multi-component formulations.

Another possibility is that products which rely on physical barriers will be developed. In British Columbia wax-based water repellents have been unsuccessfully tested for this purpose in the past but this is an area which needs to be re-examined. Additionally physical barriers might be used to encapsulate chemical or biological products on the wood surface and prevent them from being lost to the environment.

Despite the fact that biological control of biological stain is currently elusive it is possible that biological or integrated biological/chemical techniques or products can be developed.

Figures 2 and 3 illustrate recent Forintek research to seek control products of the future; the results of 1995 and 1996 testing of the biological product trade-named Cartapip. Cartapip is *Ophiostoma piliferum*, an albino mutant of a fungus implicated in stain and therefore directly occupying the ecological niche of the pest fungi. The objective of the test was to determine if the wild staining fungi, which cause discolorations, could be displaced with a white fungus which would be innocuous. The field test was done by spray applying a suspension of the fungus directly onto commercial lumber in same way chemicals are tested at Forintek (3).

The wood, 8 foot long hem-fir 2 by 4 inches, was either pasteurized or unpasteurized. Recent studies at Forintek indicate that pre-infection appears to be a fact of life in much of the hem-fir lumber being treated (4). Pasteuriza-

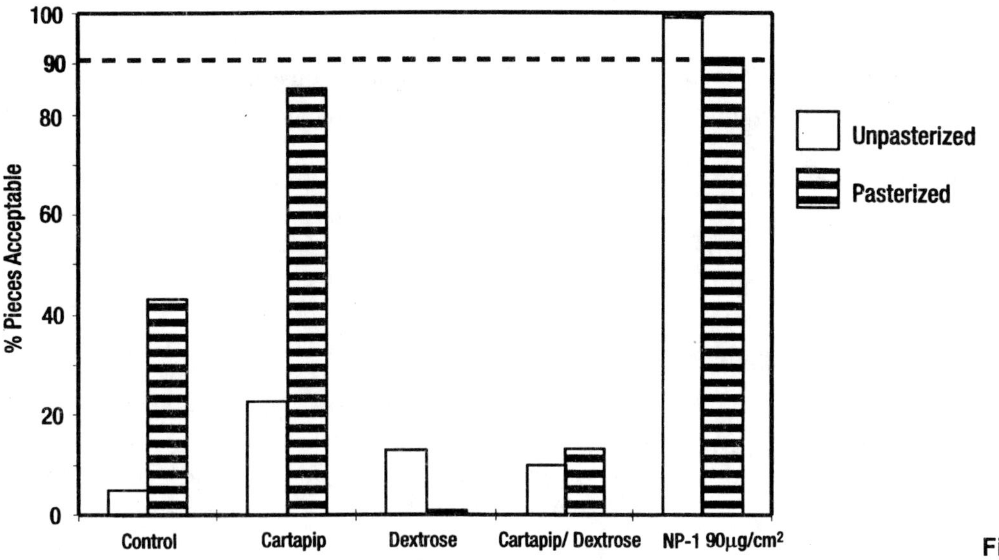

Figure 3.—1996 field test of Cartapip on pasteurized and unpasteurized wood.

tion essentially sterilizes the wood surface and is done commercially for unseasoned wood shipped to Europe ostensibly to control the pinewood nematode. We then rated the growth of fungi after 4 months of storage. Figures 2 and 3 (1995 and 1996 tests, respectively) show the percentage of the test pieces (out of 40) which were acceptable (rated 2 or less out of 5). For a treatment to be considered satisfactory 90 percent of the pieces must be acceptable.

In the 1995 test Cartapip alone on wood pasteurized or unpasteurized was not effective (Fig. 2). However Cartapip with a nutrient (5% dextrose) gave a satisfactory treatment and was as effective as NP-1 providing the wood had been pasteurized. These results gave hope that if the fungus was offered a relatively clear niche, and given a nutrient, it could work under quasi-industrial conditions.

In 1996 Forintek staff repeated the work, this time using more test parameters, for example glucose alone, sucrose as a nutrient, lower levels of inoculum, and also tested the fungus on Douglas-fir. Cartapip without nutrient was almost satisfactory on pasteurized wood but wood treated with the Cartapip/5 percent dextrose pasteurized combination (which had been successful in the 1995 trial) was badly stained (Fig. 3).

The conclusion is that under some circumstances a bio-control fungus can displace wild-type fungi and provide comparable protection to that given by chemical treatment. However it is far from certain as to what those circumstances are. The challenge is to define those conditions and to use the information to develop consistent performance. This may be particularly onerous on pre-infected wood but for protection of freshly cut logs the challenge would be significantly less and has in fact been demonstrated on logs in the United States (2).

Application Technology: Past, Present, and Future.

In the past, application of chemicals to lumber was simply a question of saturating the wood surface with chemical solution of approximately known concentration, and this was achieved by dipping whole packages in dip tanks. Unfortunately, this could result in contamination of mill yards and ground water from spillage and from dripping lumber. Although diptanks are largely regarded as past technology in western Canada a few remain there and around the world they are still a common means of applying chemicals. The few diptanks which have been installed in British Columbia in recent years have had to comply with regulations to fulfill provincial requirements and guidelines to prevent escape of chemical into the environment. Such requirements are for impervious containment areas or double-walled tanks and also for a roofed primary drip area.

To overcome problems with dip tanks, hydraulic spray systems were developed for individual board treatment in the sawmill production line. These are in common use in British Columbia and are increasingly being used in other parts of the world, such as in New Zealand and the U.S. Pacific Northwest. The earliest systems were for application of chemical when the lumber was travelling longitudinally (with the long axis pointing downline), usually after it came out of the planer. However, this meant that the treated lumber still had to be contacted by graders and greenchain workers downstream of the spray box. These workers have to be protected from chemical contact by personal protective equipment.

Transverse spray boxes, which apply chemicals to the wood when it is travelling in the transverse direction, were more recently developed to treat lumber at the packaging end of the sawmill where no further worker contact with the product would occur. However treatment with transverse boxes has been notoriously uneven because the boxes are hard to design well and maintenance requirements are high. Overall transverse spray systems are less efficient method of applying sapstain control products than longitudinal systems.

The major application requirement at present and in the future is that an even coating of chemical at particular target levels should be achieved. Even coverage gives the best chance of stain prevention and also avoids localized over treatment which is a waste of chemical and may contribute to environmental problems by dripping or leaching. Such coverage is possible with a properly set up spray system. Sprays have the potential for more even chemical coverage because the surface texture of the wood is not so critical as it is with dipping. The system can be set up to deliver the amount of chemical required and not have the uptake dictated by the surface texture of the wood as it is in dip-treatments. With sprays a treated product which is not dripping wet can be produced by putting on less liquid but using a higher concentration to achieve the same chemical retention. This minimizes worker exposure and causes less chemical to drip into the mill, yard, or surrounding environment. One other advantage of sprays is the reduction in "stripping" of active ingredients, which is a common problem in dip-tanks.

One of the issues which has arisen is that it is not always easy to optimize hydraulic spray systems. Although they are based on simple, mature technology, the interactions between wood of varying sizes, travelling at high speeds with the many forms of spray clouds obtainable are not fully understood. Forintek and others have looked at more advanced technologies such as electrostatic application and there is a considerable body of knowledge on how these perform. However, it may well be some time before this type of technology is adopted.

In the future there is little doubt that diptanks will continue to be used in some locations for lumber treatment. But dip treatment will probably not be suitable for biological protectants which will have to be sprayed. Improvements to spray technology will continue to be made for both lumber and logs. Although work has been done on electrostatic application of sapstain control products, this technology is generally not regarded as giving sufficient advantage over hydraulic spray technology to ensure its adoption in the immediate future. The application system of the future will be part of an integrated system where the process is monitored and controlled electronically. The system must be as close to foolproof as possible and of low maintenance. Whether this "smart" application system uses hydraulic or some more advanced spray technology or other techniques such as dry powder coating or foam application remains to be seen.

Monitoring Application and Quality Assurance: Past, Present, and Future

Monitoring application is done to achieve a level of quality assurance. This is based on the knowledge that a certain level of chemical (or biological should they be commercialized) retained on the lumber surface will protect lumber from stain. Although a mill can apply fungicide at an apparently known rate, it is the amount of chemical which is retained on the wood surface (chemical retention) which determines whether or not mould and staining fungi have the potential to be a problem.

In the past there was very little monitoring except for checking that a dip-tank was at the correct concentration. Methods of measuring this in the field have not proven to be very accurate. In Canada, in the late 1960s it was recognized that there was a need to check the chemical retentions being achieved on lumber (7). This developed in 1977 into the present program of routine measurement of surface retentions. Forintek had a program, in which 20 to 40 mills regularly send wood samples from their current production to the laboratory, for about 20 years. (Other mills send samples to other laboratories or to their chemical suppliers.) These are analyzed for the amount of chemical present which acts as a check of the mill's lumber protection processes and application efficiency. The results are communicated to the mill together with recommendations, where appropriate, for changes in the application of chemical. The analysis is semi-automated and data processed and faxed by a microcomputer, resulting in very fast service. The practice of sampling treated lumber surfaces has now spread to other parts of the world where it is common today.

Work at Forintek on in-line monitoring of chemical, to enable faster response to under- or over-treatment, has so far not proved successful. In the future there will probably be in-line monitoring of either chemical or biological product retentions developed. This will allow feedback to a process controller which could adjust application for changing variables, such as different wood sizes and particular market requirements, thus turning the current approach into a true system for protection of our lumber.

Summary

Through lessons learned from past practices, protection against biological stains for the present and future should be considered as a system integrating:

* An understanding of the causal organisms;
* A portfolio of control products and methods;
* Improved application technologies;
* Monitored effectiveness of treatments.

Acknowledgments

This paper is the result of the work of a small group at Forintek Canada Corp., formerly the Western Forest Products Laboratory, dedicated to minimizing discoloration problems on lumber produced by the Canadian industry. The origianl data reported here was in large part generated by David Minchin, Wood Protection Technologist at Forintek's Vancouver laboratory. Clariant Corporation, Charlotte, North Carolina, kindly supplied the cultures of Cartapip for field tests.

Literature Cited

1. American Society for Testing Materials. 1996. Standard test method for controlling sapstain and mould on unseasoned lumber (laboratory method). ASTM designation D 4445 - 1991 (reapproved 1996). Philadelphia PA. 4 p.
2. Behrendt, C.J., R.A. Blanchette, and R.L. Farrell. 1995. Biological control of blue-stain fungi in wood. Phytopathology. 85(1):92-97.
3. Byrne, A. 1994. Forintek field test method to determine efficacy of sapstain control products on unseasoned lumber. Unpublished report. Forintek Canada Corp. Vancouver B.C. 8 p.
4. Clark, J.E. 1997. Fungal growth in stored western hemlock logs. Unpublished report. Forintek Canada Corp. Vancouver, B.C. 20 p.
5. Cserjesi, A.J. and E.L. Johnson. 1982. Mold and sapstain control: laboratory and field tests of 44 fungicidal formulations. Forest Prod. J. 32(10):59-68.
6. Eades, H.W. 1956. Sap-stain and mould prevention on British Columbia softwoods. Dept. Northern Affairs and Natural Resources. Forest Products Lab. Div., Bulletin 116. Vancouver B.C.
7. Roff, J.W. and A.J. Cserjesi, 1965. Chem. Preventives used against mould and sapstain in unseasoned lumber, B.C. Lumberman 49(5):90-98.
8. Smith, R.S. 1991. Lumber Protection Today. In: Lumber Protection in the 90s. Special Publication No. SP 33. Forintek Canada Corp. Vancouver B.C. pp. 3-13.

A Microscopic Study on the Effect of IPBC/DDAC on Growth Morphology of the Sapstain Fungus *Ophiostoma Piceae*

Ying Xiao and Bernhard Kreber

Abstract

An exploratory study was performed to probe into fungal/wood/fungicide interactions using *Ophiostoma piceae*, a common sapstaining fungus on unseasoned radiata pine. Treated (1% solution containing IPBC (0.07% w/w) and DDAC (0.60% w/w)) and untreated radiata pine wafers were inoculated with *O. piceae* and examined microscopically over a 10-day incubation time.

The results of the study showed that while prolific spore germination and mycelial growth of *O. piceae* occurred on untreated radiata pine, spore deformation and lysis was observed on treated radiata pine. Using Trypan blue, a vital stain, it was found that some swelled spores remained viable during incubation for 10 days suggesting that IPBC/DDAC mode of action is to delay the process of spore germination.

Introduction

Unseasoned radiata pine (*Pinus radiata* D. Don) is highly susceptible to fungal degrade starting immediately after tree felling. In particular fungal degrade caused by sapstaining fungi represents a major problem to New Zealand's logs exporters with an estimated loss in revenue of more than $100 million annually (22). In order to control fungal degrade of logs during storage and transition to overseas markets prophylactic chemicals, referred to as anti-sapstain treatments, are applied to the wood surface either through dipping or spraying. However, current anti-sapstain formulations are unable to provide satisfactory log protection of radiata pine logs for 6 months, the time required by industry (20).

A considerable body of research is available on testing and evaluation of prophylactic treatments for controlling unseasoned wood from sapstain and mould fungi during storage and transportation (2,5,7,8,10,17,18). Generally in anti-sapstain trials the potential of a prophylactic treatment to keep unseasoned wood free of discoloration is evaluated on both the wood surface and below the wood surface (interior stain) after some storage period. However, very few studies have attempted to understand the mode of action of chemicals on wood inhabiting fungi using mainly artificial media (1,16,22) rather than wood (12,19). In general prophylactic chemicals may have fungicidal and/or sporocidal properties, that is when fungal hyphae and spores are killed, or fungistatic and/or sporostatic properties when the process of hyphal growth and spore germination is retarded (2,15). Specific information is unavailable on fungicide/wood/fungus interaction with regard to sapstaining fungi. Studying the effect of commercial and experimental anti-sapstain formulations on growth morphology may help to understand the mode of action against individual target fungi, and more specifically against sexual and asexual propagules associated with a particular target fungus. Furthermore, new anti-sapstain protection strategies may evolve from such studies.

In this preliminary study, the effect of one commercial anti-sapstain formulation containing 3-iodo-2-propynyl butyl carbamate and didecyl dimethyl ammonium chloride on growth morphology was investigated using *Ophiostoma piceae* (Münch) H. and P. Syd., a common staining fungus on unseasoned radiata pine (3,14).

Ying Xiao and Bernhard Kreber, Wood Protection Resource Center, Wood Processing Division, New Zealand, Forest Research Institute Ltd., Rotorua, New Zealand.

Material and Methods

Wood Sample Preparation and Treatment

Radiata pine (*Pinus radiata* D. Don) sapwood was sampled fresh off the saw, cut into flat-sawn (tangential) wood wafers (30 by 20 by 5 mm), sealed in plastic bags, and stored in a freezer until use. Upon thawing the wafers at room temperature, they were autoclaved (121°C, 10 minutes) and then allowed to cool down. Ten wafers selected at random were dipped in a commercial anti-sapstain formulation containing a 1 percent solution with 0.07 percent w/w 3-iodo-2-propynyl butyl carbamate (IPBC) and 0.60 percent w/w didecyl dimethyl ammonium chloride (DDAC) for 15 seconds and permitted to drain for 10 minutes. Ten additional wafers were similarly dip-treated in sterile, distilled water to serve as controls.

Inoculum preparation

Ophiostoma piceae (Münch) H. & P. Syd. was incubated on malt agar (2% malt extract, 1% agar) plates in the dark at 25°C. After 10 days of incubation 10 mL of sterile, distilled water was added to the Petri dish to flush off spores after slightly scraping the mycelial surface with a blunt glass rod. The collected suspension contained mainly conidia and very little hyphal fragments as verified microscopically. Spore density was counted using a Neubauer Chamber and yielded 2.2×10^7.

Infection and Incubation of Wafers

Treated and untreated wood wafers were infected by pipetting 1 mL of spore suspension near the edge at one side of the wafers tangential/longitudinal face (30 by 20 mm) prior to slightly tilting the wafer to allow complete wetting of the surface. After infection, wafers were placed with plastic rods into sealed glass jars with moistened filter paper on the bottom and incubated in the dark at 25°C.

Microbiological Examinations

After 1, 2, 4, 7, and 10 day(s) of incubation, microscopical examinations were performed on two wafers selected at random. Thin hand-cut sections were prepared with a sharp razor blade, stained for 10 minutes in Trypan blue, a vital stain (11). Stained and unstained sections were then examined under transmittent light using a Zeiss Photo Microscope. The same stained sections were also examined with a LEICA TCS NT Confocal Laser Scanning Microscope (CLSM). Additional hand-cut sections were immediately fixed in 6 percent glutaraldehyde for 24 hours, dehydrated by passing through a series of acetone, and then coated using a gold sputter coater. The coated samples were examined with a Cambridge Stereo Scan 240 Scanning Electron Microscope (SEM).

Results

Microscopic Observations on Untreated Wafers

Infection of untreated wafers with propagules of *O. piceae* gave rise to prolific spore germination and mycelial growth.

After 1 day of incubation with *O. piceae*, some hyphae were found at the surface of untreated wafers. In addition conidiophores with rows of conidiogenous cells arising at different levels from a single hyphae with oval shaped conidia about 3 to 5 µm long and 2 to 3 µm wide were observed (Fig. 1).

Colonization of control samples by *O. piceae* propagules was well established after 2 days showing prolific hyphal growth. Also synnemata which likely belonged to the genus *Sporothrix* spp., were found bearing

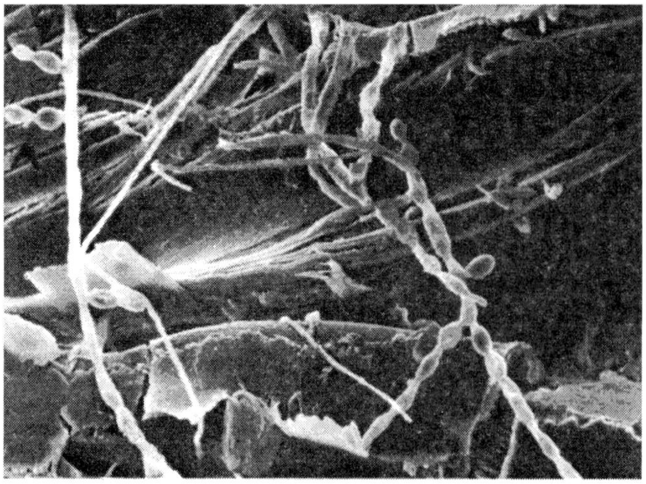

Figure 1.—Development of conidiogenous cells on untreated radiata pine after 1 day of incubation (arrow heads). SEM, 1850 x.

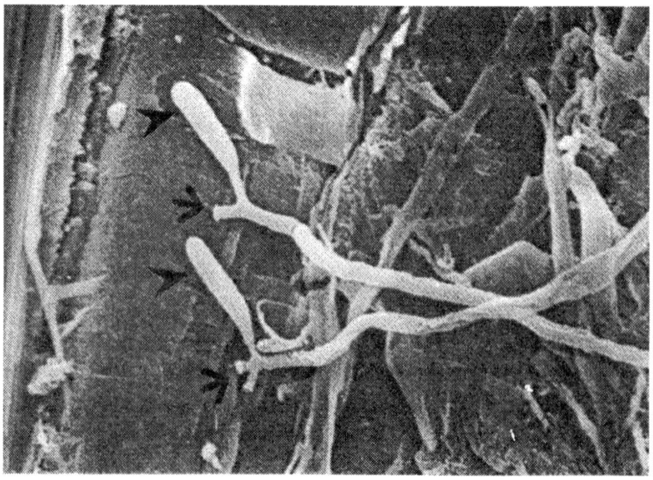

Figure 2.—*Sporothrix* sp. synnemata bearing conidia on untreated radiata pine after 2 days of incubation (arrow heads indicate conidia and arrows synnema). SEM, 5450 x.

oblong, ellipsoidal shaped conidia spores between 4 to 8 µm in length and 1 µm in width (Fig. 2).

Over the 10 days of incubation melanization of either hyphae or conidiophores did not occur on unstained, untreated radiata pine. Also characteristic *Graphium* spp. synnemata with apical conidiogenous cells born on penicillate branches were not observed.

Microscopic Observations on IPBC/DDAC Treated Wafers

Microscopic examination of chemically treated wood wafers demonstrated that propagules of *O. piceae* were unable to germinate over the 10 days of incubation.

After 2 days of incubation, deformation and lysis of conidia which had likely derived from synnemata of *Sporothrix* spp., was observed occurring on treated wafers (Fig. 3).

Figure 3.—Deformed *Sporothrix* spp. conidium on IPBC/DDAC treated radiata pine after 2 days of incubation.

After incubation for 10 days, however, some conidia had swollen to nearly round-shaped spores with a diameter of up to 7 to 8 µm (Fig. 4). These swollen conidia spores were either uninucleate or multinucleate as verified microscopically on sections stained with Trypan blue. In addition, some other slightly oval but flat-shaped conidia spores with a diameter of up to 7 to 7.7 µm were seen using light microscopy and SEM (Fig. 5). These latter conidia were not swollen and alive but may be related to the former swollen conidia after lysis had occurred. Clearly, germination of any spores did not occur on IPBC/DDAC treated wafers during the experimental period.

Discussion

The current study shows that rapid and extensive colonization and substrate occupation occurs following introduction of propagules of *O. piceae* onto untreated radiata pine sapwood. Two types of conidiophores developed on untreated wood: synnemata which were likely linked to *Sporothrix* spp. (6), and rows of conidiogenous cells arising at different levels from a single hyphae which may be related to either the *Sporothrix* spp. or *Graphium* spp. stage of *O. piceae*. However, synnemata typical of *Graphium* spp. (13) were not observed suggesting that this particular anamorphic stage may have been lost during culturing of *O. piceae* or longer incubation periods and/or conditions are needed to promote its development.

In this study, IPBC/DDAC treatment effectively inhibited growth of *O. piceae* and induced lysis of conidia derived from synnema of *Sporothrix* spp. This observation

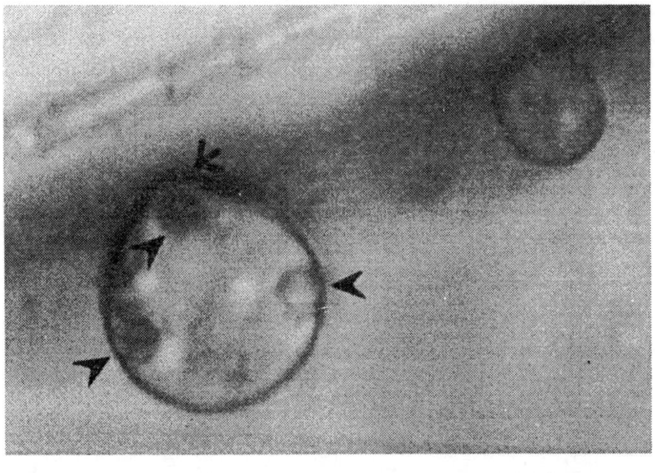

Figure 4.—Swollen, multinucleate conidia observed on IPBC/DDAC treated radiata pine after 10 days of incubation. Arrow heads indicate nuclei stained with Trypan blue and arrow possible germination site. CONFOCAL, 6800 x.

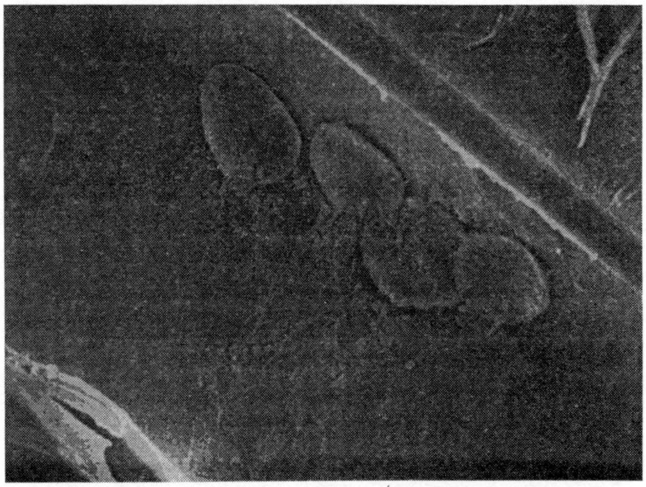

Figure 5.—Slightly oval but flat shaped conidia on IPBC/DDAC treated radiata pine after 10 days of incubation. Conidia may be derived from either *Graphium* spp. or *Sprothrix* spp. SEM, 2850 x

demonstrates that the prophylactic treatment was sporicidal (fungi toxic) and the mode of action of IPBC/DDAC possibly is due to either fungal spore membrane damage or intercellular disruption. DDAC compounds are thought to affect the semipermeable membrane of fungi causing leakage of cell constituents and also inhibition of respiratory activity (9,22). However, some other conidia spores had swollen on IPBC/DDAC treated wood, but were alive as shown microscopically with the use of Trypan blue, indicating that the treatment possibly delays the process of germination in some spores. Culturing of treated sections on nutrient media could clarify if spore germination occurs and thus the likely fungi static effect of the prophylactic treatment on these latter conidia spores.

While lysis of some *Sporothrix* spp. conidia spores occurred on treated wafers, some other conidia which may be related to either *Graphium* spp. or *Sporothrix* spp., remained viable suggesting that sensitivity of different *O. piceae* propagules towards IPBC/DDAC treatment may vary considerably as it has been reported in other related studies (12,22). Additional studies are needed to determine if spores and mycelium of *O. piceae* and its anamorphs exhibit different tolerance levels towards the IPBC/DDAC treatment. Also, tolerance level of *O. piceae* against prophylactic treatments may vary depending on the virulence of a strain and on the chemical formulation.

In general terms, the current study indicated that IPBC/DDAC varies in its activity against different *O. piceae* propugales. In addition the use of Trypan blue may be a good diagnostic tool to *in vivo* determine the effect of prophylactic wood protection treatments on fungal propagules.

Literature Cited

1. Briscoe, P.A., G.R. Williams, D.G. Anderson, and G.M. Gadd. 1990. Microbial tolerance and biodetoxification of organic and organometallic biocides. The Intern. Res. Group on Wood Preserv. Document No. IRG/WP/1464.
2. Bundgaard-Nielsen, K. and P.V. Nielsen. 1996. Fungicidal effect of 15 disinfectants against 25 fungal contaminants commonly found in bread and cheese manufacturing. Journal of Food Protection 59(3):268-275.
3. Butcher, J.A. 1968. The ecology of fungi infecting untreated sapwood of *Pinus radiata*. Canadian Journal of Botany 46:1,577-1,589.
4. Byrne, A. (ed). 1991. Lumber protection in the 90's. Proceedings Special publication No. sp. 33 (Forintek Canada Corp.), Vancouver, British Columbia (31 January). 77 p.
5. Cserjesi, A.J. and E.L. Johnson. 1982. Mold and sapstain control: laboratory andc field tests of 44 fungicidal formulations. Forest Prod. J. 32(10):59-68.
6. de Hoog, G.S. 1993. Sprothrix-like anamorphs of *Ophiostoma* species and other fungi. pp. 53-60. In: M.J. Wingfield, K.A. Seifert, and J.F. Webber. *Ceratocystis* and *Ophiostoma*. Taxonomy, Ecology and Pathogenicity. APS Press, St. Paul, Minnesota. 293 p.
7. Drysadale, J.A. and R.M.Keirle. 1986. A comparative field test on the effectiveness of antisapstain treatments on radiata pine roundwood. The Intern. Res. Group on Wood Preserv. Document No. IRG/WP/3376.
8. Drysadale, J.A., M.E. Hedley, and J.A. Butcher. 1986. An evaluation of chemical treatments for the protection of radiata pine logs from fungal degrade. The Intern. Res. Group on Wood Preserv. Document No. IRG/WP/3377
9. Eaton, R.A. and M.D.C. Hale. 1993. Wood: decay, pests and protection. Chapman and Hall, London, UK. 546 pp.
10. Eslyn, W.E. and D.L. Cassens. 1983. Laboratory evaluation of selected fungicides for control of sapstain and mold on southern pine lumber. Forest Prod. J. 33(4):65-68.
11. Gurr, 1960. Encyclopaedia of microscopic stains. Williams and Wilkins Co., Baltimore.
12. Hegarty, B. and G. Buchwald. 1988. The influence of timber species on preservative treatment on spore germination of some wood-destroying basidiomycetes. The Intern. Res. Group on Wood Preserv. Document No. IRG/WP/2300.
13. Hunt, J. 1956. Taxonomy of the genus *Ceratocystis*. Lloydia 19(1):1-58.
14. Kay, S. 1995. Control of sapstain of *Pinus radiata* with microorganisms. D. Phil. thesis. University of Auckland, New Zealand.
15. Konno, Y.T., J.O. Machado, and M.G.C. Churata-Masca. In vitro inhibition of spore germinaton of Alternaria solani by three fungicides. Revista de Microbiologia 25(1): 64-67.
16. Micales, J.A. 1990. The effect of tween 80 on the growth morphology and enzyme secretion of *Postia placenta*. The Intern. Res. Group on Wood Preserv. Document No. IRG/WP/1456.
17. Miller, D.J. and J.J. Morrell. 1989. Controlling sapstain: Trials of product group I on selected western softwoods. Forest Research Lab, Oregon Sate University, Corvallis, OR. Research Bulletin 65. 12 p.
18. Miller, D.J., J.J. Morrell, and M. Mitchoff. 1989. Controlling sapstain: Trials of product group II on selected western softwoods. Forest Research Lab, Oregon State University, Corvallis, OR. Research Bulletin 66. 10 p.
19. Sutter, H.P. and E.B.G. Jones. 1985. Interactions between copper and wood degrading fungi. Rec. Brit. Wood Pres. Assoc. Ann. Conv., 29-39.
20. Wakeling, R. 1996. Personal communication.
21. Wakeling, R. 1996. Protection of export logs from fungal degrade. Wood Processing Newsletter Issue No 19 (May).
22. Williams, G.R. and R.A. Eaton. 1988. Studies on the toxicity of biocides towards mould and sapstain fungi. 755-761p. In: D.R. Houghton and R.N. Smith (eds). Biodeterioration 7. Proceedings of the 7th Intern. Biodeterioration Symposium, Cambride, UK.

Fumigation for Preventing Non-Biological Lumber Stains

Terry L. Amburgey and Elmer L. Schmidt

Introduction

Sapwood discolorations that cannot be prevented by the application of anti-sapstain biocides occur in the lumber of many commercially important hardwood species (1). Down-grade in lumber caused by these discolorations has been estimated to cost U.S. hardwood lumber producers more than $200 million annually. These sapstains have been shown to be non-microbial and are caused by the formation of starch-like granules located in the sapwood ray parenchyma cells (5-7). It has been hypothesized that these granules are formed as a wound response and that their formation is mediated by oxidative enzymes and/or precursors located in the parenchyma cells (5-7). This hypothesis was supported when Schmidt and Amburgey demonstrated that fumigation of logs both killed parenchyma cells in sapwood and yielded lumber free of non-microbial sapstain (4,18,19). Observation from several studies indicates that sufficient starch granules only occur to be visible as sapwood discolorations when initial drying of freshly cut lumber is impeded by bulk-stacking or poor air-drying conditions (1,5-8). Non-microbial sapstains tend to be greater in lumber cut from water-stored logs than from fresh logs (9).

Prevention of Non-Microbial Sapstain

Lumber

The objective of several studies to prevent non-microbial sapstains was to "treat" freshly cut lumber to inactivate the enzymes and/or precursors mediating the formation of starch granules in parenchyma cells. Results from these studies demonstrated that discoloration could be prevented by treating lumber with a diffusible antioxidant or subjecting it to heating and/or drying by microwaves (1,5-9,23). Other studies demonstrated that these discolorations could be prevented by subjecting freshly cut lumber to mechanical stresses (3,10,13,14). The process of preventing non-microbial sapstains by mechanically stressing lumber was named the TASK Process and a U.S. patent covering this process has been allowed (2).

Logs

From earlier studies on the fumigation of oak logs, it was recognized that log fumigation may be a viable procedure for killing ray parenchyma (12,15,17). Initial trials at Mississippi State University demonstrated that short log sections of sugar hackberry (*Celtis laevigata* Wiild.) and red oak (*Quercus* spp.) fumigated for 72 hours with methyl bromide yielded lumber free of non-microbial sapwood discolorations, whereas lumber from non-fumigated logs contained these sapstains (18,19). The TTC assay indicated that the parenchyma cells in all fumigated logs were non-viable, but those in non-fumigated logs were viable (19). Other tests showed that lumber free of non-microbial sapstains could be obtained by fumigating logs of red alder (*Alnus rubra* Bong.), hickory (*Canya* spp.), and sugar maples (*Acer saccharum* Marsh.) (11,22).

Initial studies on the prevention of non-microbial sapstains indicated that these discolorations tended to be a greater problem in lumber cut from water-stored logs than from fresh logs (9). Since all previous tests had been done with lumber cut from logs directly following fumigation, Schmidt and Amburgey initiated a test to determine whether discoloration-free lumber could be obtained from logs stored under a water spray following fumigation (4). Red oak, sugar hackberry, and white ash (*Fraxinus americana* L.) logs up to 5 meters long were fumigated with one of three fumigants and then stored under a water spray with non-fumigated controls for 4 months during the

Terry L. Amburgey, Forest Products Laboratory, Forest & Wildlife Research Center, Mississippi State University, Mississippi State, Mississippi and Elmer L. Schmidt, Department of Wood & Paper Science, University of Minnesota, St. Paul, Minnesota.

summer in Mississippi. The lumber subsequently cut from the water-stored logs that had been fumigated with methyl bromide was discoloration-free (4).

Not all parenchyma were killed in logs treated with the other two fumigants, sulfuryl fluoride and metamsodium, and lumber cut from the water-stored logs had varying amounts of non-microbial sapstain. Therefore, under the conditions of this test, methyl bromide was more effective in killing parenchyma in logs than either sulfuryl fluoride or metamsodium (4). Results indicated, however, that sulfuryl fluoride may be effective in killing all parenchyma in logs at treatment levels greater than those used. This study and others (22) also indicated that growth of surface molds is greater on lumber obtained from logs fumigated with methyl bromide than on lumber from non-fumigated logs. This was in contrast to an earlier report using oak disks (16). If methyl bromide fumigation of logs is commercialized, higher concentrations of anti-sapstain biocide likely will be required to protect lumber from microbial discolorations.

The process of fumigating logs to prevent non-microbial sapwood discolorations in lumber obtained from them has been named the ESTA Process and this technology is covered in a U.S. patent (20).

Literature Cited

1. Amburgey, T.L. and L.H. Williams. 1991. Treatment of freshly sawn hardwood lumber. In: Proc., XIX Annual Hardwood Symposium of the Hardwood Research Council, Starkville, MS. pp. 105-107.
2. Amburgey, T.L. and S. Kitchens. 1995. Method for preventing and/or controlling staining in lumber, apparatus therefore and non-stained lumber. U.S. Patent Office Serial No. 08/437,371.
3. Amburgey, T.L. and S. Kitchens. 1997. Prevention of non-microbial sapwood discolorations in hardwood lumber: chemical and mechanical treatments. International Research Group on Wood Preservation.
4. Amburgey, T.L., E.L. Schmidt, and M.G. Sanders. 1996. Trials of three fumigants to prevent enzyme stain in lumber cut from water-stored hardwood logs. Forest Prod. J. 46(11/12):54-56.
5. Amburgey, T.L. and P.G. Forsyth. 1987. Prevention and control of gray stain in southern red oak sapwood. In: Proc., XV Annual Hardwood Symposium of Hardwood Research Council, Memphis, TN. pp. 92-99.
6. Forsyth, P.G. 1988. Control of non-microbial sapstains in southern red oak, hackberry, and ash lumber during air seasoning. M.S. Thesis, Department of Forest Products, Mississippi State University. 50 pp.
7. Forsyth, P.G. and T.L. Amburgey. 1992. Microscopic characterization of non-microbial gray sapstain in southern hardwood lumber. Wood and Fiber Science 23(3):376-383.
8. Forsyth, P.G. and T.L. Amburgey. 1992. Prevention of non-microbial sapstains in southern hardwoods. Forest Prod. J. 42(3):35-40.
9. Forsyth, P.G. and T.L. Amburgey. 1992. Prevention on non-microbial sapstains in water-stored oak logs. Forest Prod. J. 42(3):59-61.
10. Kitchens, S. and T.L. Amburgey. 1997. Prevention of non-microbial sapwood discolorations in hardwood lumber: chemical and mechanical treatments. In: Prevention of Discolorations in Hardwood and Softwood Logs and Lumber. Forest Products Society, Madsion, WI. pp. 26-27.
11. Kreber, B., E.L. Schmidt, and T. Byrne. 1994. Methyl bromide fumigation to control non-microbial discolorations in western hemlock and red alder. Forest Prod. J. 44(10):63-67.
12. MacDonald, W.L., E.L. Schmidt, and E.J. Harner. 1985. Methyl bromide eradication of the oak wilt fungus from red and white oak logs. Forest Prod. J. 35(7):11-16.
13. Mississippi Forest Products Laboratory. 1995. Novel methods for controlling gray stain in logs and lumber. Research Advances 4(1):1-4.
14. National Hardwood Lumber Association. 1996. Two technologies offer promising solutions to sticker stain. Hardwood Research Bulletin, No. 474. 4 pp.
15. Ruetze, M. and W. Liese. 1985. A post fumigation test (TTC) for oak logs. Holzforschung 39:327-330.
16. Schmidt, E.L. 1985. Control of mold and stain on methyl bromide fumigated red oak sapwood. Forest Prod. J. 35(2): 61-62.
17. Schmidt, E.L. 1988. An overview of the methyl bromide fumigation of oak logs intended for export to the EEC. In: Proc., Canada Wood Preserver's Association. pp. 22-27.
18. Schmidt, E.L. and T.L. Amburgey. 1993. Log fumigation prevents enzyme-mediated sapwood discolorations in hardwoods. International Research Group on Wood Preservation, Document No. 93-10003.
19. Schmidt, E.L. and T.L. Amburgey. 1994. Prevention of enzyme stain of hardwoods by log fumigation. Forest Prod. J. 44(5)32-34.
20. Schmidt, E.L. and T.L. Amburgey. 1996. Prevention of enzyme mediated discoloration of wood. U.S. Patent 5,480,679.
21. Schmidt, E.L. and T.L. Amburgey. 1997. Prevention of non-microbial enzymatic sapstain by log fumigation. In: Prevention of Discolorations in Hardwood and Softwood Logs and Lumber. Forest Products Society, Madison, WI. pp. 28-29.
22. Schmidt, E.L., D. Cassens, and J. Steen. 1997. Log fumigation prevents sticker stain and enzyme-mediated sapwood discolorations in maple and hickory lumber. In: Prevention of Discolorations in Hardwood and Softwood Logs and Lumber. Forest Products Society, Madison, WI. pp. 38-41.
23. Usman. 1994. The use of microwaves to prevent non-microbial sapstain in sugar hackberry and southern red oak. M.S. Thesis, Department of Forest Products, Mississippi State University. 30 pp.

A Study of the Factors Governing the Performance of Preservatives Used for the Prevention of Sapstain on Seasoning Wood with Regard to the Establishment of European Standards: Overview of Co-Operative Project and Development of Laboratory Test Methods

D. J. Dickinson and A. L. E. Morales

Background

The major portion of sawn timber produced in Europe is subject to sapstain and mould growth. For the production of high-quality timber this loss of aesthetic qualities is an obvious problem. However, within Europe the problem of mould and stain on lower priced package and pallet board timber currently presents an even greater problem.

In countries like Sweden producing high-quality building and furniture timber, kiln drying has expanded considerably with around 80 percent of current production being dried. Approximately 1 million cubic meters per annum of Swedish timber is still air dried and requires chemical treatment for protection during seasoning (O. Bergman, pers. comm.). Although kiln drying can prevent fungal growth, re-wetting during transport and storage and subsequent fungal development remains a problem. Wrapping of packs to prevent wetting from rain can also cause condensation making chemical protection a necessity. In Finland most timber is kiln dried but chemical treatments are still the norm to prevent stain due to this accidental wetting (7). In addition to these problems, chemical treatment is also practiced in Sweden to prevent the development of thermotolerant and thermophilic moulds during kilning which presents a major health problem in some mills (5).

In Southern Europe, air drying is still very common, and for much of the production of package and pallet boards no drying is carried out. Portugal produces 350,000 m^3 of sawn Maritime pine per annum, the majority of which is treated against sapstain (3). Effective chemical treatment is essential if, for instance, Portuguese timber is to be able to compete with other materials for pallet production and box timber. Kiln drying would not only prove prohibitively expensive but would lead to rejection of much of the timber in a market where size and distortion tolerances are critical due to automatic production of many of the final products. In South West France the production of Maritime pine depends on the widespread use of chemical treatment due to climatic conditions. Production in 1990 was 1.66 million cubic meters of sawn timber and 4.3 million cubic meters of logs (D. Dirol, pers. comm.).

In the recent past the industry worldwide has relied almost exclusively on the use of Sodium Pentachlorophenoxide (NaPCP) for sapstain control. This chemical is highly effective and very forgiving of bad application and poor quality control at the sawmills.

Sweden banned the use of NaPCP at sawmills for sapstain control in 1978 and since that date an ever increasing number of specifiers are not prepared to accept NaPCP treated timber (4). The Netherlands banned its

D. J. Dickinson, Timber Technology Research Group, Department of Biology, Imperial College, London, England and A. L. E. Morales, Graduate Student, Imperial College, London, England.

use in 1989. In France its use is still permitted but the need for change is considered urgent to allow free circulation of treated wood in Europe. Irrespective of legislation, (e.g., 9th Amendment directive 76/769/CEE (1991)) customer pressure is such that alternative, more environmentally acceptable treatments are being demanded throughout Europe. As a result there has been worldwide interest in the development of NaPCP replacements. Within Europe this has been pioneered in Sweden (5) and Finland (7), but recently a major effort has been concentrated in Portugal where UK, Dutch, and German interests are demanding pallet and packaging timber free of NaPCP.

The development of safer anti-stain chemicals and their assessment has been frustrated by the lack of established standard methods of test, making it impossible to compare published information. As a result it was necessary to carry out extensive testing in any potential new market. This lack of standard procedures has also led to new products entering the market on the strength of manufacturers claims, only to give problems in use. National requirements are also being set up to police the chemical treatment carried out on treated timber in countries importing the timber.

The European Committee for Standardisation (CEN) Technical Committee 38 (TC38) requested that the subject of sapstain testing be placed on their program. It was therefore very urgent to complete this work so that CEN could approach their work from a sound scientific basis after this prenormative work.

Co-Operative Project

The work within this project was funded with the European 3rd Framework for funding Scientific Research available for bids between 1991 and 1994. The current project fell within the AAIR program covering Agriculture, Agro-Industries including Fisheries. The work was on a shared cost basis where the European community provides 50 percent of the funding of the total program. There were nine participants in the project:

1. Imperial College of Science, Technology & Medicine, United Kingdom.
2. Centre Technique Du Bois Et De L'Ameublement, France.
3. TNO Building and Construction Research, The Netherlands.
4. The Swedish University of Agricultural Sciences, Sweden.
5. Forest Authority UK, United Kingdom.
6. Laboratorio Nacional De Engenharia Civil, Portugal.
7. Institut fur Holzbiologie und Holzschutz, Germany.
8. Consorcio De Industriais Exportadores De Madeiras, LDA, Portugal.
9. Instituto Nacional De Investigacion Y Tec-Nologia Agraria Y Alimentaria, Spain.

The project ran for 40 months and finished in March 1997.

The project consisted of a range of core tasks and related sub-tasks. This paper briefly describes the nature of the whole project but specifically details the work directed to establish a laboratory procedure.

The scientific objectives of the project were: to define the limits of any possible standard test designed to assess the preventative action of anti-sapstain chemical treatments in Europe, to establish the scientific basis for preparing such standards as requested by CEN.TC.38, and also to investigate novel non-toxic treatments in a separate study. This will make it possible to establish laboratory and field test methods relevant to the different parts of the European community and the different wood species grown therein.

Core Tasks in the Project

The first part of the project consisted of three linked core tasks, essential sub-tasks and associated sub-tasks related to the core project and of significant interest in specific regions of the European Community. The three core tasks set out to establish the limits of any proposed standard test approach, and to demonstrate any relationships that exist between any proposed laboratory assessment, field test, and actual practice. The core tasks take into account the variables that exist within the community with regard to climatic factors, wood species, saw mill practice, and causal organisms. A fourth task was concerned with novel treatments and was conducted in Sweden.

Task 1 and its associated sub-tasks consisted of laboratory studies with materials supplied from the different geographical regions of the community. The work established the basis for acceptable and meaningful laboratory procedures on which a future standard could be based. Details of this work are presented in this paper.

Task 2 and its associated sub-tasks extended the study to field trials in different parts of the community with different, economically important softwoods. Details of this work are presented in this proceedings (6).

Task 3 put into practice the findings that could be drawn from Tasks 1 and 2. This was essentially to validate or not the procedures and establish the limits of any future standards. This work will be published at a later date.

Task 4 was conducted in Sweden and was concerned with non-chemical control methods. Details of the task are presented in this proceedings (2).

The ultimate objective of the work was to make recommendations to CEN TC 38 regarding the suitability and nature of European Norms designed to assess the effectiveness of treatments for the prevention of sapstain and mould growth in green, sawn timber.

Details of Task 1: The Development of Laboratory Test Methods

Introduction

The principal tasks in this part of the project were designed to develop a meaningful, reproducible laboratory procedure. The trials were based on well established procedures using miniboards' of green test timbers treated with a range of selected, representative chemicals. The boards were treated with mixed spore inocula of a range of test fungi supplied by the various participants. A full combination of variables with a range of fungicides, wood species, and fungi from different geographical regions of Europe were examined. The object was to establish the relationships between these variables in order to make a universally acceptable recommendation for a laboratory test method. Several other factors of importance arose during the work and were studied in order to strengthen the recommendations. The main task of developing a laboratory procedure was studied in a series of experiments, loosely described in the main laboratory task as "wood fungal interactions" leading to defining a specific laboratory protocol.

The principal experiments conducted in this part of the project consisted of the following.

Initial trials to compare wood species susceptibility.—Trials were conducted to investigate softwood species variation and the response of mould and stain fungi when four timber species were treated with three anti-sapstain formulations.

Establishing the concentration ranges of the test chemicals.—Trials were conducted using Corsican pine sapwood and Host fungi only. A wide concentration range for the six anti-sapstain chemicals was used. The aim was to establish treatment concentration ranges for the six selected fungicides and to identify any problems with the chosen experimental methods. This work established the basis for further work and development of the test method.

Full wood species comparative trials with moulds and stains.—An extensive series of laboratory trials were performed to determine the effect of any wood fungus interaction dependant on timber species and the source of the colonizing fungi. This was the main body of the work and fundamental to establishing an acceptable, universal approach to the selection of test timber and fungi in any proposed standard.

Euromould and Eurostain comparative trials.—Based on the above interaction trials, common sets of selected mixed fungal inocula were compared to each of the six timbers' Host fungal sets for each of the timbers, with the aim of establishing selected Euro sets of fungi equal to or more aggressive than the Host colonizing fungi.

Comparison of laboratory trials to field trials.—Finally the results obtained from the laboratory investigations were compared to the results obtained in the field by the other participants in the project.

Materials and Methods: The Establishment of Laboratory Test Methodologies

Almost all the work in this part of the project was performed according to protocol agreed upon at the initial meetings, and the details evolved during the research once several parameters had been investigated.

General Procedures and Materials

Timber.—Each participant was responsible for ensuring a regular supply of their local pine species. These consisted of: French Scots pine, English Corsican pine, Swedish Scots pine, Spanish Radiata pine, Portuguese Maritime pine, and French Maritime pine.

Test blocks consisted of sapwood measuring 40 by 40 by 8 mm with open faces being of radial longitudinal orientation. Prepared wood blocks were sterilized by gamma irradiation and stored frozen for up to 1 year.

In the initial wood species trials UK grown timber was used for all wood species until supplies from the other participants became available.

Fungi.—Formulations for testing were challenged with two mixed spore suspensions of selected mould or sapstaining fungi. In the initial trials this was conducted with the local fungal species supplied by the participants. Each partner conducting field trials supplied five principal mould and stain fungi from their local pine species. These sets of fungi are referred to as Host sets and relate directly to the local wood species from which they were isolated. In later trials a common set of mould or stain was selected from these fungi and referred to as the Euromould or Eurostain set.

Test chemicals.—The chemicals chosen for all the comparative laboratory and field trials were selected from:

copper-8-quinolinolate, sodium pentachlorophenate, chlorothalonil, and tri-methyl cocoammonium chloride and 2 ethyl hexanoate (an AAC mixture). Formulations containing Propiconazole and 2-(thiocyanomethylthio)-benzothiazole (TCMTB) were also used in several laboratory trials.

The concentrations used were normally $2x, x, \frac{1}{2}x$, and $\frac{1}{4}x$ where x equals the manufacturers suggested effective concentration. The specific chemicals were supplied as Active Ingredients or as products, but the products were not identified.

Treatment.—The blocks were weighed, treated by dipping in solution for 10 seconds, drained, and weighed again. There were five replicate blocks per concentration, and the treated blocks were placed on racks in incubation chambers over water.

Inoculation.—The blocks were sprayed with a spore suspension using an airbrush. Suspensions were made by scraping the surface of established cultures in 10 mls of sterile Aerosol-OT solution. This surfactant was made up by using 0.05 g of the sodium salt of dioctyl sulphosuccinate per 100 g of distilled water. The resulting spore counts were between 1×10^6 and 1×10^8 per cm^3.

Incubation.—The racks were stored at 100 percent relative humidity and at 25°C for 2 weeks.

Assessment.—The extent of fungal colonization was normally evaluated according to the following scheme.

Visual assessment grading categories:

0. No growth, board is clean.
1. 0 to 25% coverage. Traces of visible fungi apparent.
2. 25% to 50% coverage. Fungi established.
3. 50% to 75% coverage. Sporulation becoming heavy.
4. 75% to 100% coverage. Full growth, underlying wood obscured.

Specific Materials and Methods

Timber.—English grown Scots, Corsican and Radiata pine, and Norway spruce. These timbers were harvested in southern England.

Chemicals.—The selected fungicidal formulations were copper-8-quinolinolate, propiconazole, and chlorothalonil. The concentrations used for all three chemicals were 1 percent, 0.5 percent, 0.25 percent, and 0.125 percent (w/w).

Fungi.—Leptographium lundbergii and *Ceratocystis adiposa* (Stains) concentration 6.75×10^6 spores per ml, and *Trichoderma harzianum* and *Penicillium funiculosum* (Moulds) concentration 4.8×10^7 spores per ml were used. UK grown timber was used in these early trials until supplies from Europe became available.

Establishing the concentration ranges of the test chemicals.—The experiment utilized freshly felled, irradiated Corsican pine sapwood and tested all six test chemical. The treatment solutions consisted of seven concentrations for each chemical prepared by serial dilution (by a factor of 2) of a 6 percent stock solution. The freshly prepared, mixed spore suspensions used were: mould species from English Corsican pine: 3.76×10^{10} spores per liter. Staining species from English Corsican pine: 1.005×10^{11} spores per liter.

The mould fungi were identified as: *Alternaria* sp., *Sphaeropsis sapinea*, *Trichoderma* sp., *Mucor* sp., and *Cladosporium cladosporoides*. The staining fungi were identified as: *Ceratocystis coerulescens*, *Leptographium wingfieldii*, *Leptographium truncatum*, *Leptographium procurum*, and *Ophiostoma piceae*.

Full wood species comparative trial with moulds and stains.—This experiment used all six wood types and all six sets of the Host mould and staining fungi isolated from them. Each set of fungi was used to inoculate all the different timbers supplied from the different countries in Europe. This resulted in 72 separate combinations of timber and fungi. A single chemical, copper-8-quinolinolate, was used in this work at two concentration ranges: 3 percent, 1.5 percent, 0.75 percent and 0.37 percent for the mould fungal sets and 2 percent, 1 percent, 0.5 percent, and 0.25 percent for the staining fungal sets.

Euromould and Eurostain Comparative Trials

The Euromould set.—This experiment used all six European pine species which were treated with four chemicals: copper-8-quinolinolate, sodium pentachlorophenate, chlorothalonil, and the AAC mixture at four concentrations:

Copper-8-quinolinolate: 3 percent, 1.5 percent, 0.75 percent, and 0.37 percent.

AAC Mixture: 6 percent, 2 percent, 0.66 percent, and 0.22 percent.

Chlorothalonil: 3 percent, 1 percent, 0.33 percent, and 0.11 percent.

Sodium pentachlorophenate 3 percent, 1.5 percent, 0.75 percent, and 0.37 percent.

The wood species were inoculated with the original set of Host mould fungi and a second set inoculated with the selected Euromould set which consisted of: *Trichoderma* spp. isolated from English Corsican pine; *Trichothecium roseum* isolated from Spanish Radiata pine; *Cladosporium cladosporoides* isolated from English Corsican pine; *As-*

pergillus niger isolated from Swedish Scots pine; and *Penicillium* spp. isolated from French Scots pine.

The Eurostain set.—The experiment used all six European pine species and was treated with copper-8-quinolinolate and chlorothalonil at the following concentrations: 1 percent, 0.5 percent, 0.25 percent, and 0.12 percent. The wood was inoculated with the Host sapstaining fungi and a second set was inoculated with the selected Eurostain set which consisted of: *Ophiostoma piceae* isolated from English Corsican pine; *Leptographium procurum* isolated from English Corsican pine; *Ceratocystis coerulescens* isolated from Swedish Scots pine; LNEC 4 isolated from Portuguese Maritime pine; 35A isolated from Spanish Radiata pine; and PM10 isolated from French Maritime pine. (Fungal cultures with code numbers have not yet been identified).

Comparison of laboratory trials to field trials.—Laboratory tests were compared to field tests by selecting a pair of equivalent trials, determining the concentration of chemical which gave a Visual Assessment Grading of 1, recording the two corresponding concentrations and dividing the highest by the lowest. The number arrived at is called the multiplication factor and represents the degree of difference between the two concentrations that give equivalent fungal growth in the two different tests.

For example, if a laboratory and a field trial have identical growth responses to the same concentration the resultant number would be 1. If, however, the field trial required twice the concentration to give an equivalent degree of control this number would be 2. The multiplication factor calculated will have a value greater than 1 in favor of the test with the higher concentration, that test being the most challenging of the two. Comparisons of the concentrations are made at the Visual Assessment Grade of 1 because this value corresponds most closely to the Minimum Inhibition Concentration, which is that concentration which just fails to control fungal growth and is therefore considered suitable for making these comparisons.

Table 1.—Order of susceptibility (rated 1 to 4).

Inoculum	Chemical formulation	Scots pine	Corsican pine	Radiata pine	Norway spruce
Moulds	Chlorothl.	2	1	3	4
Moulds	Cu-8	1	2	4	3
Moulds	Prop	1	2	3	4
Stains	Chlorothl.	1	2	3	4
Stains	Cu-8	1	2	4	3
Stains	Prop.	1	3	2	4

Results

Initial Trials to Compare Wood Species Susceptibility

The trial was conducted to investigate softwood species variation with regard to the response to mould and stain fungi when four woods were treated with three anti-sapstain formulations. It was found that the susceptibility of the wood species could be ranked (according to the vigor of fungal growth) at any given concentration of chemical. The relative rating of the wood species was only marginally influenced by which chemical or fungal inoculum was used.

The results showed that generally Scots pine was the most susceptible with Corsican and Radiata pine following in that order (Table 1). Norway spruce was the least susceptible. This observation was confirmed in all subsequent trials where different wood species were used.

The most sensitive wood species was consistently Scots pine with Corsican pine in a few cases. The least sensitive to colonization was Spruce. Radiata pine was the least susceptible pine species and on occasions showed less colonization than the Spruce. This ranking was based on the comparison of the visual assessment means of the four wood species for each treatment. Statistical analysis was performed by arranging the data into groupings for each concentration and for each fungal inoculum, allowing the timber species and the chemical to be compared. The mean, sum, and sum of squares for the data allowed for an Analysis of Variance to be carried out, following which a t test could be performed to compare the difference between the two means to the corresponding value found on the table of Least Significant Differences. From this work the significance of the difference between the two means in question could be ascertained, therefore validating the timber susceptibility ranking.

Establishing the Concentration Ranges of the Test Chemicals

The trial was conducted using Corsican pine sapwood and its Host fungi. A wide concentration range for the six anti-sapstain chemicals was used. Growth of dominant fungi was observed with changing concentration in the stain set.

Mould and stain data sets gave satisfactory results except for the TCMTB stain and Na-PCP mould sets which had either zero or seriously limited growth. This is thought to be due to a vapor effect which was investigated separately. In further tests and in the protocol it is recommended that different concentrations of chemical are incubated in separate chambers to avoid any effects caused by volatile compounds.

Working concentrations that provided a full range of fungal response were selected for further laboratory use

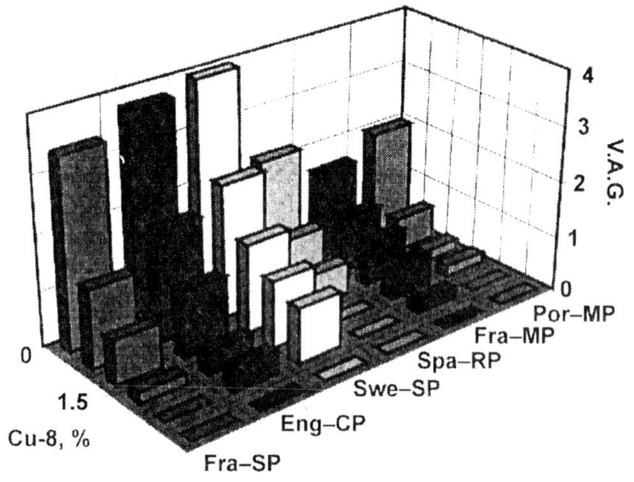

Figure 1.—All six pines inoculated with vigourous mould inoculum: moulds isolated from French Scots pine. V.A.G. visual assessment grading; Cu-8 % concentration of copper-8-quinolinolate, %; Fra-SP French Scots pine; Eng-CP English Corsican pine; Swe- SP Swedish Scots pine; Spa-RP Spanish Radiata pine; Fra-MP French Maritime pine; Por-MP Portuguese Maritime pine.

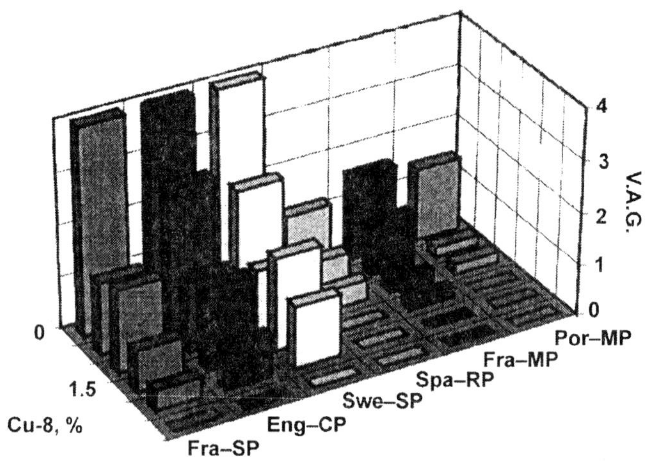

Figure 2.—All six pines inoculated with less vigourous mould inoculum: moulds isolated from French Maritime pine. V.A.G. visual assessment grading; Cu-8 % concentration of copper-8-quinolinolate, %; Fra-SP French Scots pine; Eng-CP English Corsican pine; Swe- SP Swedish Scots pine; Spa-RP Spanish Radiata pine; Fra-MP French Maritime pine; Por-MP Portuguese Maritime pine.

from the results of the large, initial range (Table 2). These concentrations give the full range of fungal growth responses on all six wood species with either mixed mould or mixed stain fungal inocula. In this way, other parameters affecting the response of the technique can be tested for reliably.

In further trials a simple disposable incubation rack and chamber were used to avoid the observed effects of volatile compounds.

Full Wood Species Comparative Trials with Moulds and Stains

The full range of comparisons between wood species and source of fungal inoculum was performed in the laboratory with copper-8-quinolinolate. From these results it was shown that timber species was the overriding factor in test design, as previously shown in the initial comparative trial. The timbers were placed into susceptible and less susceptible groupings, based on the degree of colonization, at a given concentration, of either the mould or stain inoculum. These groupings were:

Group 1 (susceptible): Scots and Corsican pine.

Group 2 (less susceptible): Radiata and Maritime pine.

The fungi could also be placed into similar groups (vigorous and less vigorous) based on the timber they were isolated from. The mould (and to a lesser extent stain) groups showed a lack of Host wood species specificity. The less vigorous staining sets showed a degree of specificity to their Host timbers (notably Maritime pine) but this observation had little relevance in selecting a staining group for the testing of new formulations. These findings were more closely examined in the Euromould and Eurostain experiments.

Figure 1 shows the Group 1 timber susceptibility with moulds isolated from French Scots pine. The first 3 columns from left to right, represent the Visual Assessment Grade Averages (V.A.G.s) for the Scots and Corsican pine timbers. For any given concentration it can be seen that this V.A.G.s are correspondingly higher than the results for the Radiata and Maritime pine, represented by the last 3 columns. This indicates that these Group 1

Table 2.—Selected working concentration ranges.

Chemical	Mould				Stain			
	%				%			
Na-PCP	0.4	0.45	1.5	3	0.06	0.25	1	4
Copper-8	0.4	0.75	1.5	3	0.25	0.5	1	2
Chlo-rothl.	0.1	0.3	1	3	0.12	0.5	2	8
Propi-cozl.	0.1	0.3	1	3	0.1	0.3	1	3
AAC	0.2	0.6	2	6	0.2	0.6	2	6
TCMTB	0.06	0.25	1	4	0.1	0.3	1	3

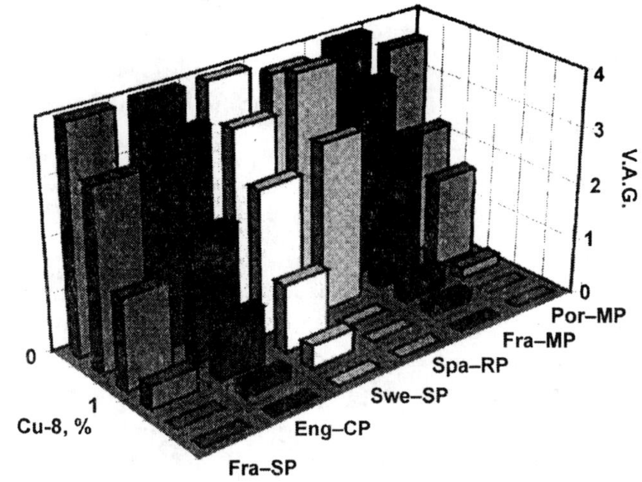

Figure 3.—All six pines inoculated with less vigourous stain inoculum: stains isolated from Spanish Radiata pine. V.A.G. visual assessment grading; Cu-8 % concentration of copper-8-quinolinolate, %; Fra-SP French Scots pine; Eng-CP English Corsican pine; Swe-SP Swedish Scots pine; Spa-RP Spanish Radiata pine; Fra-MP French Maritime pine; Por-MP Portuguese Maritime pine.

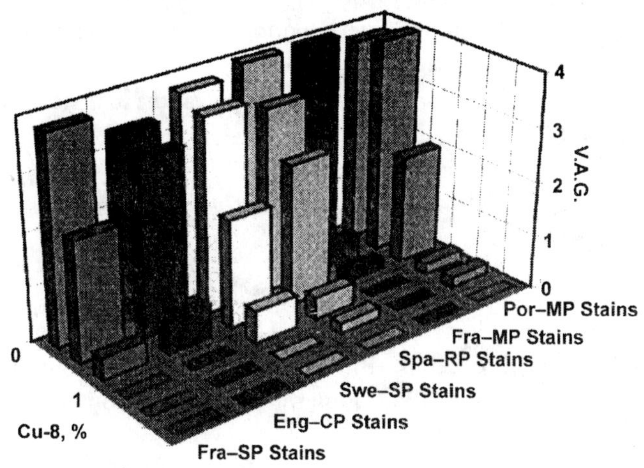

Figure 4.—Less susceptible pine inoculated with all six stains: French maritime pine timber. V.A.G. visual assessment grading; Cu-8 % concentration of copper-8-quinolinolate, %; Fra-SP French Scots pine stains; Eng-CP English Corsican pine stains; Swe-SP Swedish Scots pine stains; Spa-RP Spanish Radiata pine stains; Por-MP Portuguese Maritime pine stains; Fra-MP French Maritime pine stains.

timbers are more susceptible to the colonization and growth of moulds isolated from French Scots pine.

Figure 2 shows a similar result but with moulds isolated from French Maritime pine. Again, the Group 1 timbers are the most susceptible even though the inoculum is from Group 2 timber, i.e., French Maritime pine. This result is important as it shows that timber susceptibility is a more important factor than the source of the inoculum.

Figure 3 shows similar results are obtained with staining fungi (from Spanish Radiata pines). This indicates that when a less vigorous staining group is used to inoculate all the pines they respond similarly.

Figure 4 shows a degree of specificity in the Maritime pine stain inoculum. The timber in question was French Maritime pine inoculated with stains from Portuguese Maritime pine (6th column). The Portuguese Maritime pine inoculum grows well in comparison to the fungi isolated from the Group 1 timbers (first 3 columns), but exceeds the growth of French Maritime pine inoculum on its Host timber (5th column).

Euromould and Eurostain Comparative Trials

These trials were performed to establish selected Euro sets of fungi that were equal to or more aggressive than the Host fungi isolated from the different European pine species. Within this experiment it was also possible to validate the observations made in the previous section. The experiment was designed so that a direct comparison could be made between a Euro set of fungi and the fungal set originally isolated from the timber species being challenged, referred to as the Host fungi. It was therefore possible to utilize the means from the Visual Assessment Grades to perform an Analysis of Variance, per concentration, to determine how significant the difference was between the means for Euro set fungi and host fungi. As this method uses data collected from all six timber species, it is also possible to use the results generated from the analysis of variance to investigate how significant the differences were between the two postulated timber species groupings observed in the previous section.

Euromould set results.—The visual assessment results were used to compare the two sets of mould fungi (i.e., Host set against selected Euro set) for each pine species. There were 96 statistical comparisons, from which 38 were found to be significantly different. From these, 12 differences were found to be due to a stronger response by the Host fungi, of which four were only at the 10 percent level. Therefore, out of 96 comparisons made between the Euromould set and Host mould fungi, 88 comparisons were insignificantly different or in favor of the Euromould set, i.e., the Euromould set gave a response equal to or stronger than the Host mould set.

Eurostain set results.—The visual assessment results for fungal colonization were used to compare the two staining sets (i.e., Host set against selected Euro set) for each pine species. There were 72 statistical comparisons, from which 33 were found to be significantly different. From these, five differences were found to be due to a stronger response by the Host fungi. Therefore, out of 72 comparisons made between the Eurostain set and Host staining fungi, 67 comparisons were insignificantly different or in favor of the Eurostain set, i.e., the Eurostain set gave a response equal to or stronger than the Host staining set.

Timber species susceptibility (Group 1 & Group 2) results.—Group 1 timber species (Scots pine and Corsican pine) were also compared to Group 2 timber species (Radiata pine and Maritime pine) in the Euromould and Eurostain trials to determine the ranking of their susceptibilities.

The visual assessment results for the trials were used, as before, to carry out a series of comparisons.

For the 540 comparisons made with the Euromoulds and Host fungi, 25 (4.63%) were significantly different in favor of Group 2 timber species being more susceptible to colonization than the Group 1 wood species.

For the 360 comparisons made with the Eurostains and Host fungi, 37 (10.3%) were significantly different in favor of the Group 2 timber species being more susceptible to colonization than the Group 1 wood species.

The great majority of these comparisons back up the observation that Group 1 timbers are more susceptible and so may be recommended for use in future trials to provide a severe challenge to fungal control.

The comparisons were made across the full range of concentrations for the statistical analysis. The results were also checked at the point of failure (which corresponds approximately to the Visual Assessment Grade of 1), and it was found that the selection of the Group 1 timber species group was particularly valid at this point. The selection of the Euro sets of fungi were also found to be valid at this point.

Comparison of Laboratory Trials to Field Trials

The laboratory and field trials were compared by using the multiplication factor method previously described. The Minimum Inhibition Concentration (corresponding to a Visual Assessment Grading of 1) was used as a measure of the difference between the two situations. The laboratory trials were compared with field trials using the same wood species and chemical and using the most severe degree of fungal colonization (i.e., either the Euromould or Eurostain set). The field trials were compared once the packs treated with 1.5 percent copper-8-quinolinolate had failed.

Out of 47 valid comparisons made, 32 (68%) were less than twice as different (had multiplication factors of 2 or less). Of the remaining 15 (32%) comparisons, 14 were in favor of the more severe situation in the field.

Therefore, it can be stated that the majority of laboratory trials are comparable to field trials. Where they were not comparable, they are almost always less severe, i.e., they fail-safe in comparison to the field trial, which is considered to be desirable in any laboratory test of this type.

Discussion

The diversity of organisms and wood species made it difficult to suggest a single standard approach to laboratory testing of anti-sapstain chemicals. The wood/fungal interaction experiments proved very complicated, but in the presence of fungicides the origin of the fungal species did not seem to be of overriding importance. Probably the most important findings were with regard to the wood susceptibility. It seems clear that Scots and Corsican pine are more prone to fungal growth than the other species, when tested under identical conditions. Certain compounds showed effects due to volatiles and different concentrations should be tested separately.

The use of a common Euro mould and stain set of fungi and the choice of the most susceptible timber species, namely Scots or Corsican pine, made it possible to postulate a standard laboratory protocol. This is based on comparison to an internal fungicidal standard of copper-8-quinolinolate. The results obtained from this method also compare favorably to full scale field trials.

The following recommendations and proposals have been forwarded to CEN.TC.38 to form the basis of a definitive European standard test method.

Outline Proposal for a Standard Laboratory Protocol for the Testing of Proprietary Anti-Sapstain Chemicals

Protocol Requirements: Procedures and Materials

Timber.—Freshly sawn, flawless sapwood of susceptible pine species to be used (either Scots pine or Corsican pine). The size should be 40 by 40 by 8 mm with open faces being of radial longitudinal section. Prepared wood blocks should be irradiated. Blocks may be cold stored temporarily or refrigerated for up to 1 year.

Fungi.—Formulations for testing are to be challenged with 2 separate, mixed spore suspensions of selected European mould and sapstaining fungi. These fungi are:

Euromould set.—Trichoderma sp. isolated from English Corsican pine; *Cladosporium cladosporoides* isolated from English Corsican Pine; *Penicillium* sp. isolated from French Scots Pine; *Aspergillus niger* isolated from Swedish Scots Pine; and *Trichothecium roseum* isolated from Spanish Radiata Pine.

Eurostain set.—Ophiostoma piceae isolated from English Corsican Pine; *Leptographium procurum* isolated from English Corsican Pine; *Ceratocystis coerulescens* isolated from Swedish Scots Pine; LNEC 4 isolated from Portuguese Maritime Pine; 35A isolated from Spanish Radiata Pine; and PM10 isolated from French Maritime Pine.

Test chemicals.—Internal standard: Copper-8-quinolinolate. Concentrations: 1.5 percent and 0.75 percent. Test chemical: four concentrations, $2x$, x, $\frac{1}{2}x$, and $\frac{1}{4}x$ where x equals the manufacturers suggested effective concentration. A control set of wood blocks dipped in distilled water should be included.

Treatment.—Blocks should be weighed, treated by dipping in solution for 10 seconds, drained, and weighed again. Five replicate blocks should be allowed per concentration. Treated blocks are then placed on racks. Separate incubation chambers are used for each chemical and concentration. Blocks are conditioned for 24 hours after treatment.

Inoculation.—Blocks are sprayed with a spore suspension using an airbrush. Suspensions are made by scraping the surface of established cultures in 10 mls of sterile Aerosol-OT solution. Spore counts should be between 1×10^6 and 1×10^8 per cm^3.

Incubation.—Racks are stored at 100 percent relative humidity and at 25°C for 2 weeks.

Assessment.—The extent of fungal colonization is evaluated according to the following scheme. Visual assessment grading categories:

0. No growth, board is clean.
1. 0 to 25% coverage. Traces of visible fungi apparent.
2. 25% to 50% coverage. Fungi established.
3. 50% to 75% coverage. Sporulation becoming heavy.
4. 75% to 100% coverage. Full growth, underlying wood obscured.

Literature Cited

1. Anonymous. 1991. Council Directive (91/173/EEC) amending for the Ninth time Directive 76/769/EEC on the approximation of the laws, regulations and administrative provisions of the Member States relating to restrictions on the marketing and use of certain dangerous substances and preparations. Official Journal of the European Communities, OJ No L/85, 5.4.91, pp 34-36.
2. Bjurman, J. 1998. Novel non-toxic treatments for sapstain control. In: Biology and Prevention of Sapstain. Forest Products Society, Madison, WI. pp. 93-99.
3. Dickinson, D.J. 1987. Sapstain control on Portuguese Maritime pine imported into the U.K. Proceedings of the I.U.F.R.O. Wood Protection subject Group S. 5. 03.
4. Henningsson, B. 1977. Health and safety aspects of the use of wood preservatives in Sweden. The International Research Group on Wood Preservation, Document IRG/WP/396.
5. Henningsson, B. 1987. Prevention of stain and mould on sawn softwoods. The situation in Sweden. Proceedings of the I.U.F.R.O. Wood Protection subject Group S. 5. 03.
6. Nunes, L. and D.J. Dickinson. 1998. European collaborative field trials. In: Biology and Prevention of Sapstain. Forest Products Society, Madison, WI. pp 87-92.
7. Vihavainen, T. 1987. Prevention of stain and mould on sawn softwoods. The situation in Finland. Proceedings of the I.U.F.R.O. Wood Protection subject Group S. 5. 03.

European Collaborative Field Trials

Lina Nunes and David Dickinson

Abstract

The objective of this study was to define the limits of any possible standard field test designed to assess the preventive action of anti-sapstain chemical treatments in Europe. In order to do so several drafts were developed which led the participants of the project to a final document with general recommendations to the European Committee for Standardisation, Technical Committee 38 (Durability of wood and wood products) regarding the implementation of a future standard.

Introduction

The existence of several non-standardized test methods and procedures in Europe and in the world in general, concerning sapstain tests and the introduction of new anti-stain products determined the need to establish the scientific basis for a standardized field test method of these kind of wood preservatives. This test would assess the effectiveness of treatments for the prevention of sapstain and mould in green, sawn timber.

In January 1994, a collaborative project involving nine organizations of seven different countries, was initiated within the framework of the Agriculture and Agro-industry, including the Fisheries program (AIR). The project entitled "A study of the factors governing the performance of preservatives used for the prevention of sapstain on seasoning wood with regard to the establishment of European standards" had the following objectives:

* To define the limits of possible standard tests designed to assess the preventive action of anti-sapstain chemical treatments in Europe.
* To establish the scientific basis for preparing such standards as requested by CEN/TC 38 in October 1992.

Lina Nunes, Timber Division, Laboratório Nacional de Engenharia Civil, Portugal and David Dickinson, Timber Technology Research Group, Department of Biology, Imperial College, London, England.

The project consisted of three linked core tasks set out to establish the limits of any proposed standard test approach and to demonstrate the relationships that exist between any proposed laboratory assessment, field test, and actual practice.

The work described here refers to the establishment of the basis for acceptable meaningful field test procedures on which a future standard could be based. In order to do so, a protocol for a field test method was devised and tests were conducted, twice in 1 year (starting in spring and in autumn), in five countries with four pine species. After the completion of all programmed field tests and evaluation of the results, the general recommendations regarding the implementation of a future Standard by CEN.TC 38 were agreed upon.

Materials and Methods

Materials

A first draft of the test method was drawn based on the experience of the different participants involved in the project and also on relevant national standard methods already in existence (1-3). Six sites were chosen to conduct parallel field tests while considering the different climatic regions of the community and the local pine species (Table 1). Differences in sawmill practice, for example open- and close-stacked storage, were also taken into consideration.

For each product, concentration and condition of exposure, including reference preservative and untreated controls, 55 boards were used. The dimensions of the boards varied within pre-established limits and are presented in Table 1.

As for wood preservatives, the products and concentrations shown in Table 2 were used. Sodium pentaclorophenate (Mitrol G-ST Beads) was initially taken as the reference product but it was not included in the final recommendations. A copper-8-quinolinolate wood preservative was shown to be also a convenient reference.

Table 1.—Species of wood used, location of the field tests, and test board dimensions.

Wood species	Location	Participant	Length	Width	Thickness
			mm		
Corsican pine (*Pinus nigra*)	Exeter - UK	FA	950	100	25
Maritime pine (*Pinus pinaster*)	Linxe - France	CTBA	900	100	20 to 22
	Leça da Palmeira - Portugal	LNEC	1000	100	25
Radiata pine (*Pinus radiata*)	Vasque country - Spain	INIA	1100	100	20
Scots pine (*Pinus sylvestris*)	Fontaine le Port - France	CTBA	1000	100	20 to 26
	Uppsala - Sweden	SUAS	1100	100	25

FA = Forest Authority; CTBA = Centre Technique du Bois et de l'Ameublement; LNEC = Laboratório Nacional de Engenharia Civil; INIA = Instituto Nacional de Investigación y Tecnología Agraria y Alimentaria; SUAS = Swedish University of Agricultural Sciences

Table 2.—Concentration of products tested.

Concentraion of products tested (%)				
Mitrol GS beads	PQ8	Sinesto B Basiment SB	Basiment 545[a]	TuffBrite[b]
0.75	0.75	2	1	1
1.5	1.5	4	2	2
3	3	8	4	4

[a] Exceptions: Sweden (both tests); Portugal and UK (spring/summer test)
[b] Exception: France (both tests)

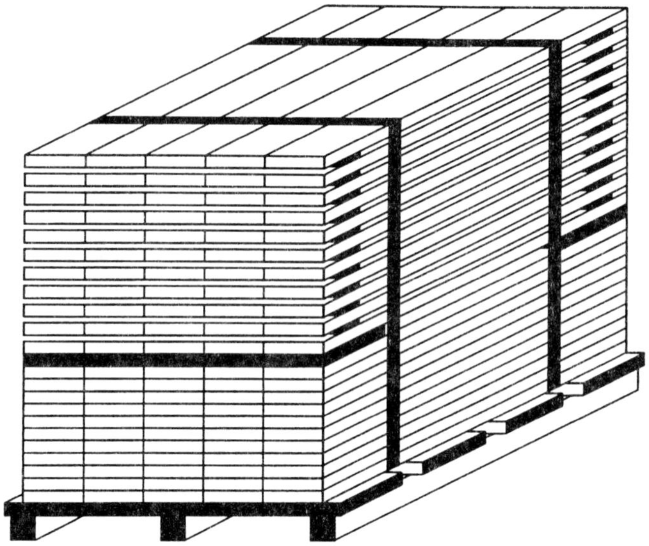

Figure 1.—Example of storage conditions.

Methods

The general experimental conditions were similar to the ones described in the proposal for field test methodology presented below. The storage of the boards is shown in Figure 1.

The tests were initiated in the spring and in the autumn. The spring-summer trials were set in April/May 1994 in Portugal, Sweden, and the UK and in April/May 1995 in France and Spain. All autumn-winter tests were set in September/October 1994. Meteorological data from all test sites have been obtained and the average values of temperature (°C), precipitation (mm), and relative humidity (%) are presented in Figures 2 to 4.

The assessment of the attack by the micro-organisms involved was done after 3 and 6 months of exposure and followed the rating method described in the proposed procedure.

Results and Discussion

The general objective of this study was to identify relevant parameters to consider when outlining a possible standard field test designed to assess the preventive action of anti-sapstain chemical treatments in Europe. Within that scope several tests were conducted and the average results obtained for all test sites and products are presented in Table 3.

The field tests conducted were conclusive as regarding the need for such a standard since some previously disregarded parameters were found to be relevant and not completely answered by laboratory testing only. One possible example is the different behavior of open- and close-stacked material in the field.

Agreement could be reached in what concerned methods and materials as well as in the general design of the test procedure. Validation criteria was adopted for both untreated and reference preservative treated material. The performance of the reference preservative was included as a positive control to ensure that a high level of risk has been present throughout the test, since the untreated controls will tend to fail even when the conditions are relatively mild and the treated material will not be conveniently stressed.

The tests have been initially considered to be valid whenever the following conditions were met: a) the control untreated boards (close-stacked material) must have at least a mean rating of 3.5; b) the reference preservative treated boards (close-stacked material) must have at least a mean rating of 1 for the 1.5 percent concentration. The

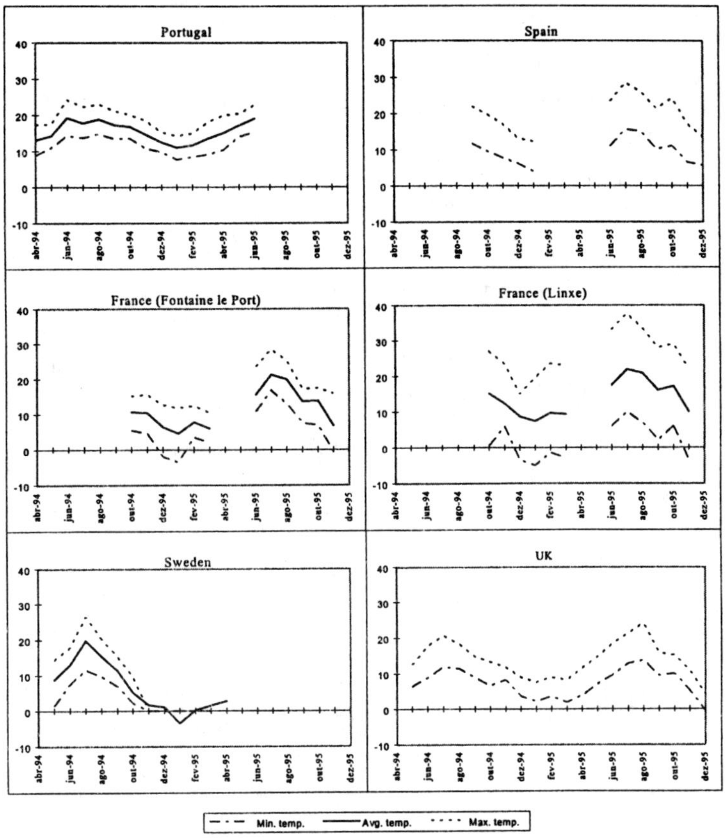

Figure 2.—Minimum, maximum, and average temperature (°C) in the different sites during the complete period of tests.

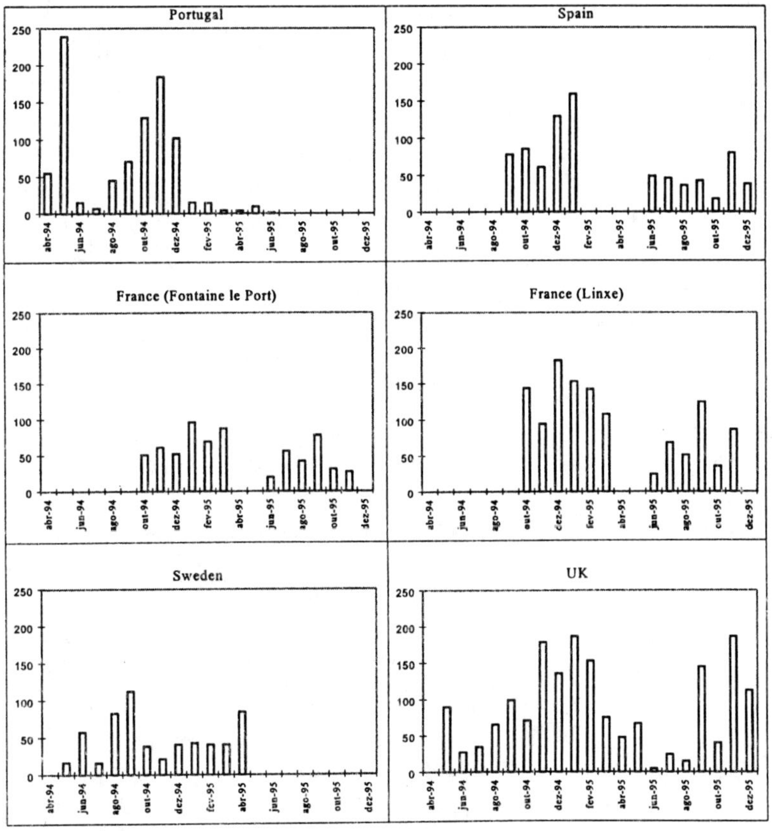

Figure 3.—Total precipitation per month (mm/m^2) in the different sites during the complete period of tests.

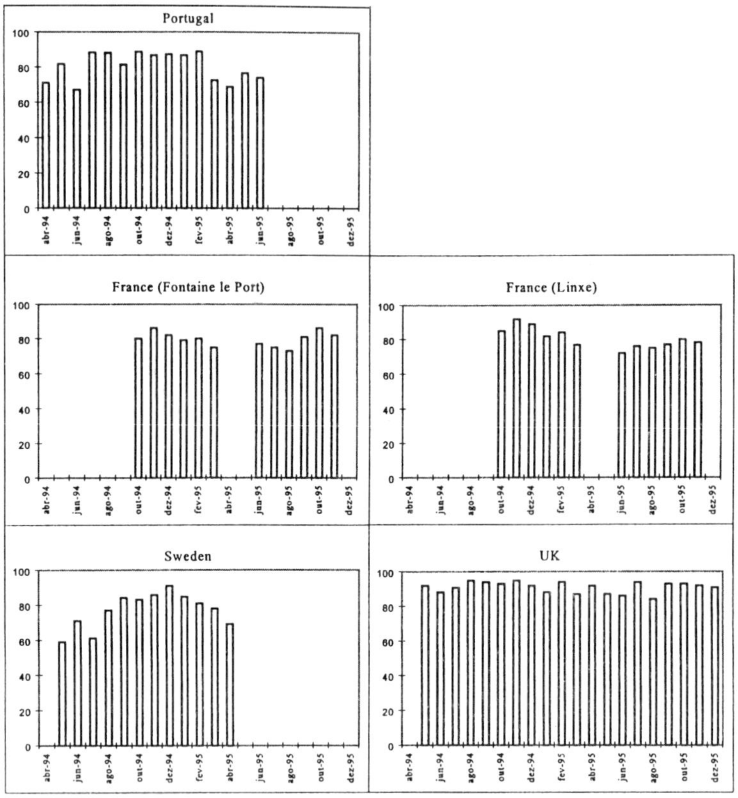

Figure 4.—Average relative humidity (%) in the different sites during the complete period of tests.

test results, however, showed that this procedure should be improved. The first finding was that the relative performance of different preservative treatments could vary depending on whether close-stacked or open-stacked material was tested; therefore close-stacked material results could not be extrapolated and both situations should always be tested. Secondly, the use of PCP Na has been restricted by recent legislation on chemical products.

From the other products included in the trial tests, PQ8 was considered a suitable alternative reference preservative since there is already a reasonable history of its use in the industry, its formulation is well known, and it presents good test results. Moreover, in the 1.5 percent concentration its performance is equivalent to PCP Na in the same concentration, as confirmed by the mean rating of 1 evaluated in the treated boards. Besides, it was decided to adopt this concentration (1.5%) since the 0.75 percent concentration also studied fails in all situations, and the higher concentration (3%) is too good to be used for the purpose of a reference preservative.

Consequently, the initial conditions for validation were revised and the outcoming proposal is that the test is valid if the following conditions are met: a) the control untreated boards (close-stacked material) must have at least a mean rating of 3.5; b) the control untreated boards (open-stacked material) must have at least a mean rating of 2.5; c) the reference preservative treated boards (close-stacked material) must have at least a mean rating of 1 for the 1.5 percent concentration; and d) the reference preservative (PQ8) treated boards (open-stacked material) must have at least a mean rating of 0.75 for the 1.5 percent concentration.

Having established this acceptance criteria, and having in mind the corresponding mean rating, some of the trials here described cannot be considered as valid, namely in the case of the open-stacked material subject to the spring/summer tests that took place in Spain and Portugal and the autumn/winter tests in Spain and Sweden. For the close-stacked material, the spring/summer test in Spain and the autumn/winter tests in France and Sweden should also be considered as not valid. When looking at the these failures, and considering the meteorological data presented in Figures 2 to 4, it appears that either extreme hot or extreme cold weather may negatively affect test results. It can thus be concluded that test program carried out in different locations or under different atmospheric conditions may not be comparable. Laboratories in charge of carrying out such tests must be aware of this variable and plan the test accordingly.

As suspected after the first set of field trials were completed, it was observed in parallel work (4) that the height of the stacks have an effect in colonization, there-

Table 3.—Comparative results between countries.

Open-Stacked Material

Preservative	Conc. % w/w	Spring/Summer Portugal MP	Spring/Summer Spain RP	Spring/Summer Sweden SP	Spring/Summer France SP	Spring/Summer France MP	Spring/Summer UK CP	Autumn/Winter Portugal MP	Autumn/Winter Spain RP	Autumn/Winter Sweden SP	Autumn/Winter France SP	Autumn/Winter France MP	Autumn/Winter UK CP
PCP Na	0.75	1.90	0.00	2.38	2.96	2.08	3.87	3.42	0.36	0.08	0.52	1.04	2.93
	1.5	0.86	0.00	1.40	2.11	2.80	3.60	0.58	0.20	0.02	0.24	0.30	1.38
	3	0.24	0.00	0.50	1.42	1.48	2.49	0.42	0.13	0.00	0.02	0.02	0.87
PQ 8	0.75	1.42	0.02	1.90	2.32	1.40	3.91	1.94	0.00	0.64	1.02	1.38	2.93
	1.5	0.32	0.00	1.58	2.55	1.45	3.93	1.66	0.00	0.18	0.95	0.78	1.62
	3	0.11	0.00	1.02	0.82	1.10	2.84	1.40	0.00	0.02	0.58	0.32	1.42
Tuff Brite	1	0.44	0.27	0.94			3.53	2.90	1.84	0.68			2.98
	2	0.32	0.20	0.78			3.44	3.08	1.48	0.92			3.36
	4	0.30	0.16	0.60			2.42	2.06	1.76	0.30			3.56
Sinesto B	2	3.10	0.11	2.02	2.23	2.87	3.96	3.40	0.48	1.34	3.60	3.94	4.00
	4	1.66	0.04	1.82	2.82	1.98	3.98	2.78	0.22	0.20	1.00	1.56	3.69
	8	0.90	0.02	1.30	1.45	1.34	3.69	2.86	0.00	0.00	0.40	1.20	2.69
Basiment 545	1		0.00		0.76	0.54		2.64	0.12		0.00	0.14	2.76
	2		0.00		0.48	0.02		2.18	0.04		0.00	0.06	1.62
	4		0.00		0.14	0.00		1.30	0.00		0.02	0.02	0.60
Untreated		3.96	0.95	2.48	3.60	3.82	4.00	4.00	3.82	4.00	4.00	4.00	4.00

Close-Stacked Material

Preservative	Conc. % w/w	Spring/Summer Portugal MP	Spring/Summer Spain RP	Spring/Summer Sweden SP	Spring/Summer France SP	Spring/Summer France MP	Spring/Summer UK CP	Autumn/Winter Portugal MP	Autumn/Winter Spain RP	Autumn/Winter Sweden SP	Autumn/Winter France SP	Autumn/Winter France MP	Autumn/Winter UK CP
PCP Na	0.75	1.76	0.78	3.62	3.06	3.46	3.93	3.36	1.55	0.34	2.48	3.94	3.51
	1.5	0.78	1.13	3.22	3.30	2.88	3.36	2.38	1.60	0.18	1.98	2.74	2.22
	3	0.18	0.36	1.10	0.72	1.24	3.33	1.98	1.43	0.10	0.66	0.30	1.87
PQ 8	0.75	3.02	0.75	2.70	3.30	3.04	3.91	3.66	1.26	0.30	0.63	1.42	2.33
	1.5	2.40	0.60	2.20	3.08	1.48	3.93	3.54	1.38	0.00	0.90	1.04	1.91
	3	0.92	0.07	0.36	1.64	0.92	3.20	1.18	1.23	0.04	0.17	0.44	1.24
Tuff Brite	1	1.82	2.05	3.04			4.00	3.80	3.30	0.90			1.16
	2	2.40	2.69	2.98			4.00	3.40	4.00	0.76			0.87
	4	2.26	2.61	2.80			3.96	3.32	3.41	0.32			3.49
Sinesto B	2	3.72	1.44	3.28	3.78	3.40	4.00	4.00	3.29	1.06	2.92	4.00	4.00
	4	2.64	2.80	1.56	4.00	3.16	4.00	3.96	3.47	0.28	1.09	3.98	3.69
	8	2.34	1.84	0.92	2.62	1.42	4.00	4.00	2.68	0.04	0.08	3.84	2.33
Basiment 545	1		1.80		2.76	1.22		3.60	1.67		0.80	1.04	1.82
	2		2.51		1.00	0.10		3.18	1.89		0.12	0.02	0.96
	4		1.77		0.60	0.42		2.22	0.77		0.04	0.00	0.82
Untreated		4.00	3.93	4.00	4.00	4.00	4.00	4.00	4.00	4.00	4.00	4.00	4.00

fore the test protocol was amended to clearly state that the trial should be set up as a monolayer.

Moreover, from the results obtained in the field trials no relation was found between the average recorded uptake of preservative and the average results obtained. One participant (CTBA) carried out specific uptake studies with PQ8 on two wood species (Scots and Maritime pine) and with Basiment 545 on Maritime pine (5). The results showed that for a certain dipping time, there is no relationship between surface retention and either the wood species or surface roughness of the boards when prepared with two different blades. Furthermore, dipping time from 30 seconds to 3 minutes appears to have no influence on retention.

Finally, other studies conducted within the project proved that Corsican and Scots pine are the most susceptible of the studied species, thus pointing out that these species should generally be considered when designing field test programs (6). However, there is evidence that every wood species always presents higher susceptibility to the indigenous fungi. Therefore, the use of commercially important local species (e.g., Maritime pine in Southern Europe) should always be adopted for these tests.

Final Proposal for Field Test Methodology

Wood

Wood species: Corsican or Scots pine (or Maritime pine, where that species is known to be susceptible).

Wood quality: Freshly cut timber, up to 1 week after cutting of the log and less than 24 hours after sawing the boards.

Dimensions of the boards: Length: at least 1,000 mm; width: 100 to 120 mm; thickness: 20 to 25 mm.

Number of boards: Close-stacked storage: 55 boards; Open-stacked storage: 55 boards.

Wood Preservative

Reference preservative: Copper-8-quinolinolate wood preservative at two concentrations: 1.5 percent and 3 percent.

Test products: A concentration range including at least one failing concentration.

Method

Time of the year: Spring or Autumn.

Treatment: Dipping, one by one, for 15 seconds.

Uptake of wood preservative: Recommended to record.

Exposure: The packages should be made according to Figure 1 and should be placed in the field on a single layer.

Meteorological data: Recommended to record.

Assessment: After 3 months of exposure, then monthly until validation criteria is met to a maximum of 6 months.

Rating: 0 = clean (0% disfigurement); 1 = slight (< 10% disfigurement); 2 = medium (10% to 25% disfigurement); 3 = heavy (25% to 50% disfigurement); 4 = severe (> 50% disfigurement).

Conditions of the validity of the test: Open-stacked material: a) the control untreated boards must have at least a mean rating of 2.5; b) the reference preservative treated boards must have at least a mean rating of 0.75 for the 1.5 percent concentration.
Close-stacked material: a) the control untreated boards must have at least a mean rating of 3.5; b) the reference preservative treated boards must have at least a mean rating of 1 for the 1.5 percent concentration.

Acknowledgments

The authors would like to thank the Commission of the European Communities for financial support of this work as well as the other participants in the project, namely, Mrs. Françoise Thomassin (CTBA, France), Dr. Rolf-Dieter Peek (BFH, Germany), Dr. Petra Esser (TNO, The Netherlands), Dr. Teresa de Troya (INIA, Spain), Mr. Osten Bergman (SUAS, Sweden), and Dr. Joan Webber (FC, UK),

Literature Cited

1. Association française de normalisation. 1993. Produits de préservation du bois. Détermination de l'efficacité fongicide des produits de protection temporaire des sciages frais. Méthode sur site. NF X41-549. AFNOR, Paris, France.
2. Eidgnenossische Materialprufungs-und Forschungsanstalt. 1992. Holzschutzmittel. Bestimmung der fungiziden Wirkung gegen Primärbläue und Schimmelbefall an saftfrischem Holz im Freiland-Versuch. EMPA - Methode - BS 3. Auflage. EMPA, St. Gallen, Switzerland.
3. European and Mediterranean Plant Protection Organization. 1995. Guideline for the efficacy evalution of fungicides. Bluestain of softwood. EPPO Bulletin 25:527-536.
4. Dickinson, D.J. 1996. AIR Project 1059. Periodic Report (January - July, 1996). Imperial College of Science, Technology and Medicine. London.
5. Dickinson, D.J. 1995. AIR Project 1059. Periodic Report (January - July, 1995). Imperial College of Science, Technology and Medicine. London.
6. Williams, J.R., D.J. Dickinson, and J.F. Webber. 1997. The effect of stack height on the performance of preservatives used for the prevention of sapstain on seasoning wood. The Int. Res. Group on Wood Preserv. Doc. no IRG/WP97-10192. IRG Secretariat, Stockholm.

Novel Non-Toxic Treatments for Sapstain Control

Jonny Bjurman, Björn Henningsson, and Hans Lundström

Abstract

The work presented is an overview of findings within part of a EU project on sapstain control dealing with novel treatments. The following potential non-toxic treatments for sapstain control were studied within the project: bioprotectants, melanin formation inhibitors, stabilization of high pH at the wood surface, and electricity. Several new or previously tried organisms including a bacterium, an actinomycete, yeast fungi, other fungi, lichens, and combinations of bioprotectants and boric acid were tried for inhibitory effects against bluestain fungi. Particularly the inhibitory effects of *Streptomyces griseoviridis* and lichens are interesting and should be studied further. The effect of compounds which had previously been reported to inhibit the formation of melanin in certain fungi were tried against bluestain fungi. The melanin formation of *Aureobasidium pullulans* was shown to be less influenced by the inhibitors than was earlier studied fungi. In field tests stabilization of high pH at the wood surface was largely unsuccessful probably due to a too rapid decrease of the pH. Applying an electric field was effective against the growth of bluestain fungi. The results are briefly discussed and some future prospects are given.

Introduction

Bluestain fungi cause a reduction in the value of timber or timber products by discoloring sapwood, thereby decreasing the aesthetic quality of wood. However, the attack of bluestain fungi might also cause several other unfavorable effects which decrease the value of the wood notably a reduction in pulp quality, increased permeability, and increased susceptibility to attack by decay fungi of untreated and painted wood (8,9).

Jonny Bjurman, Björn Henningsson, and Hans Lundström, The Swedish University of Agricultural Sciences, Department of Forest Products, Uppsala, Sweden.

Traditionally fungicidal treatments have been used to protect the wood from bluestain attack during storage, drying, and transport. However, the percentage of bluestain attack might be considerably decreased by the use of sound handling of the wood and by decreasing the time the wood is stored before drying (19). During storage of logs before drying, prevention of colonization of bluestain fungi is often accomplished by sprinkling the logs with water. In the last decades the use of kiln drying has expanded substantially. In Sweden around 90 percent of the total sawmill production is kiln dried. Although these measures have led to a decreased need for anti-sapstain treatments, an increased concern of the possible detrimental effects of biocides to the environment has encouraged the development of alternative measures or treatments for prevention of sapstain.

Among possible alternatives to conventional chemical treatments, the use of bioprotectants has been the subject of great interest and much study. A large number of micro-organisms have been tried for potential biological control and many have also been tried for prevention of blue stain including Bacteria (34) *Bacillus* sp. (5,20,25), *Pseudomonas* sp. (6,25,33); Actiomycetes *Streptomyces* sp. (25); Yeasts (34) *Kloeckera apiculata* (29), *Candida guillermondi* (29); and other fungi *Ophiostoma piliferum* (12), nonpigmented Ophiostoma - Cartapip (17), *Trichoderma* sp. (14,25), *Debaryomyces hansenii* (31), *Gliocladium* sp. (25), *Acremonium breve* (22), and *Scytalidium* (24). Integrated biological-chemical treatments have also been tried (2).

Due to its chemical binding characteristics melanin binds copper and other positively charged compounds and might accordingly decrease the effects of fungicides. Inhibition of melanin formation might therefore result in inhibition at lower concentrations of conventional fungicides. A decreased melanin content in bluestain fungal hyphae might be achieved by the use of melanin formation inhibitors (3).

In this review we present the findings from part of a European project which focused on characterization and development of novel non-toxic methods for sapstain control. Among possible alternatives to traditional fungicidal treatments earlier proposed and studied, methods such as biological control, the use of melanin inhibitors and pH control, as well as new methods such as treatment with electricity, was studied. Based on new and earlier findings the potential and future possibilities for sapstain control with alternative treatments are briefly discussed.

Materials and Methods

Bioprotectants

Within the project a large number of experiments were conducted to study the effects of various antagonistic organisms on several sapstain fungi. Direct influence of the antagonists on growth and spore germination as well as the influence of various culture filtrates and extracts from antagonists were studied. A test panel consisting of four to six common sapstain fungi was used (Table 1) in the tests. Several different bioassays employing various test substrates or growth media were also used: agar, solid wood, nutrient solutions. Antagonistic organisms which were tried included the yeast *Candida kefyr* isolated from yoghurt, an actinomycete *Streptomyces griseoviridis*, two basidiomycetes, *Bjerkandera adusta* and *Gliocladium* sp., and a light colored *Ophiostoma* sold under the commercial name Cartapip. These organisms were all previously known for antagonistic effects against fungi but not tried against sapstain fungi.

Earlier studies have shown that extracts from some lichens reduce or even prevent growth of certain decay fungi (18,27). Lichens and extracts from lichens were therefore also studied with respect to inhibitory effects against bluestain fungi. Different bioassays with different types of application of the extracts were tried which give some indications of the mode of action of the bioprotectants.

Since borates are regarded as acceptable from the environmental point of view we have also studied the combined growth inhibiting effect on sapstain fungi of very small amounts of boric acid and antagonists or their culture filtrates.

Table 1.—Mold and stain fungi which were used to evaluate bioprotection.

Aureobasidium pullulans (de Bary) Arnaud
Ceratocystis piceae (Münch) Bakshi
Ceratocystis pilifera (Fr.) c.Moreau
Cladosporium herbarum (Pers.) Link
*Hormonema dematioides**
Scopularia phycomyces (Anersw.) Goid

*Taxonomically closely related to *A. pullulans*.

Melanin Formation Inhibitors

Within the project three compounds reported to possess melanin synthesis inhibitor activity have been tested in surface liquid cultures for their effect on melanization in *Aureobasidium pullulans* and *Sclerophoma pityophila*, Coumarin, Cerulinin, and Phtalide (3).

Inhibitory Effects of Electricity

Within the project a series of experiments was made to study the effects of electricity, including laboratory experiments on both agar substrate and wood. *Aureobasidium pullulans* and *Ceratocystis piceae* were selected as test organisms. A more detailed description of these studies has been given earlier (10).

Control of High pH at the Surface of Wood

The inhibitory effect of stabilization of high pH at the wood surface on the growth of bluestain fungi was studied. Based on earlier experiences from laboratory studies (30) treatment with sodium metaborate, which initially stabilizes a high pH (over 10) at the wood surface, was selected for field trials.

Results and Discussion

Bioprotection Organisms

In the experiments with an array of potential bioprotectants a great variation in the sensitivity of the different tested bluestain fungi to the different bioprotectants was revealed. It might therefore be necessary to use a combination of bioprotectants for prevention of blue stain development.

The yeast *Candida kefyr* showed a clear growth inhibiting effect on *Aureobasidium pullulans* when the organisms were co-cultured. However, sterile filtrated solutions from liquid yeast cultures were not effective in any of the tested quantities.

The white rot fungus *Bjerkadera adusta* has earlier been shown to have a parasitic behavior against bluestain fungi (7). The present experiments verified that new strains of *B. adusta* inhibited, grew over, and caused lysis of hyphae of *A. pullulans*. However, culture filtrates of *B. adusta* did not influence growth of *A. pullulans* or *Hormoderma dematioides* but initiated pigmentation in *A. pullulans*. The actinomycte *Streptomyces griseoviridis* is used commercially against fungal diseases in tomato and cucumber plants. The commercial product is named Mycostop. When grown on the same medium *S. griseoviridis* almost completely inhibited growth of *A. pullulans*. Also culture filtrates of *S. griseoviridis* strongly inhibited the

Table 2.—Effects of antagonists in combination with boric acid on the growth of bluestain fungi. Growth of four bluestain fungi on malt agar media containing different concentrations of boric acid and filter sterilized culture solutions from liquid cultures of *Streptomyces griseo-viridis, Ophiostoma piliferum*, and *Gliocladium viride*. Linear growth of the bluestain fungi was measured after 14 (*Streptomyces*) and 12 days (*Ophiostoma, Gliocladium*), respectively.

Antagonist vs bluestain fungus	Culture filtrate conc. % (v/v)	Growth (mm) at various boric acid concentrations (% w/v) 0.001	0.025	0.05	0.25	0.5	1.0	Contr.
Streptomyces griseo-viridis								
A. pullulans	5				18	21	22	
	10				15	17	19	37
	20				7	8	10	
C. piceae	5	30	23	12				
	10	24	22	9				37
	20	18	18	8				
C. herbarum	5				34	34	34	
	10				25	29	27	49
	20				28	30	30	
S. phycomyces	5	31	29	2				
	10	30	28	3				40
	20	22	20	5				
Ophiostoma piliferum								
A. pullulans	5				30	30	30	
	10				29	30	30	32
	20				31	30	30	
C. piceae	5	31	30	1				
	10	28	29	0				33
	20	30	30	0				
C. herbarum	5				45	44	46	
	10				46	47	45	49
	20				46	45	45	
S. phycomyces	5	24	26	1				
	10	25	27	0				34
	20	23	23	0				
Gliocladium viride								
A. pullulans	5				30	30	31	
	10				30	29	30	32
	20				30	31	30	
C. piceae	5	34	30	9				
	10	34	31	10				33
	20	33	30	12				
C. herbarum	5	45	44	44				
	10	44	45	44				49
	20	43	45	44				
S. phycomyces	5	32	34	1				
	10	29	28	1				34
	20	29	21	1				

growth of *A. pullulans*. *Ceratocystis* was somewhat less sensitive.

When culture filtrates from *Bjerkandera adusta, Gliocladium viride, Candida kefyr*, and *S. griseoviridis* were added to the growth medium in combination with low amounts of boric acid various interferences occurred between boric acid and the antagonists as regards the effect on the growth of the bluestain fungi. Results for *S. griseoviridis, O. piliferum*, and *G. viride* are given in Table 2. The antagonists and the bluestain fungi were differentially sensitive to boric acid (Table 3). *C. herbarum* was resistant to both boric acid and the tested antagonists. However, none of the tested combinations and concentrations had a generally strong growth reducing effect on all bluestain fungi involved. Also usnic acid, a common lichen metabolite demonstrated growth limiting effects (Table 4). Experiments with lichen extracts were also studied. The effects of the four lichens *Cladonia alpestris, C. rangiferina, C. silvatica*, and *Hyphogymnea physodes* on the four bluestain fungi *A. pullulans, C. piceae, H. dematioides*, and *S. phycomyces* were studied (Table 5). *H. physodes* was the most active of the lichens tested. Growth on malt extract agar with different combinations of boric acid and ground thallus of *H. physodes* also revealed a great difference in the sensitivity of four tested bluestain fungi (Table 6).

According to previous studies a number of different types of agents produced by the bioprotective organisms are responsible for the fungicidal or growth inhibiting

Table 3.—Effects of boric acid on antagonists and bluestain fungi. Results after 9 days of incubation are shown.

	Boric acid concentration (%) 0.0	0.1	0.25	0.5	1.0	2.0
	- - - - - - - - Mycelial growth in mm - - - - - - - - - -					
Hormonema dematioides	21	18	17	14	7	3
Bjerkandera adusta	39	9	0	0	0	0
Ceratocystis piceae	28	0	0	0	0	0
Cladosporium herbarum	20*	25	24	24	26	20**
Scopularia phycomyces	30	0	0	0	0	0
Aureobasidium pullulans	27	21	22	21	17	12
Gliocladium viride	>65	20	0	0	0	0
Candida kefyr	+	+	+	0	0	0
Streptomyces griseoviridis	+	+	0	0	0	0

* dark colony; **light, transparent colony

For non-mycelial organisms: ++ = normal growth; + = weak growth; 0 = no growth.

Table 4.—Growth effects of usnic acid on four different blue stain fungi.

Bluestain fungus	Growth (mm) after 9, 13, and 20 days at different amounts of usnic acid*														
	0	1	10	50	100	0	1	10	50	100	0	1	10	50	100
	- - - - - - - - - - 9 days - - - - - - - - - -					- - - - - - - - - - 13 days - - - - - - - - - -					- - - - - - - - - - 20 days - - - - - - - - - -				
A. pullulans	12	10	11	9**	3	45+	45+	45+	45+	10	45+	45+	45+	45+	27
C. piceae	11	12	10	7	5	45+	45+	45+	45+	12	45+	45+	45+	45+	32
H. demat.	12	12	11	11	6	19	20	20	19	12	37	37	37	35	25
S. phycomyc.	3	0	0	0	0	7	2	0	0	0	31	30	30	30	30

* Round filter papers were soaked in acetone solutions of usnic acid of the following concentrations: 0.0%, 0.001%, 0.005%, 0.01%, and 0.05%. After drying the filter papers were placed on malt agar plates and inoculated with agar mycelium pieces of the blue stain fungi.

** Underlined figures represents clear growth reducing effects.

+ means that the mycelium had grown out of the filter paper.

Table 5.—Growth effects of ground thalli or hot water extracts from ground thalli of the lichens *Cladonia alpestris, Cladonia rangiferina, Cladonia silvatica,* and *Hypogymnea physodes.*

Treatment	Growth (% of the control) on malt agar plates with various concentrations of ground lichens or hot water extracts from lichens							
Fungus species	- - - *C. alpestris.* - - -		- - - *C. rangiferina*- - -		- - - *C. silvatica*- - -		- - - *H. physodes*- - -	
Ground lichens	0.5%	1.0%	0.5%	1.0%	0.5%	1.0%	0.5%	1.0%
A. pullulans	68*	72	80	84	84	72	40	40
C. piceae	81	76	110	110	95	86	27	7
H. dematioides	75	67	79	83	96	83	54	15
S. phycomyces	58	68	100	110	142	121	157	110
Hot water extracts	25%	50%	25%	50%	25%	50%	25%	50%
A. pullulans	59	50	105	86	91	86	105	90
C. piceae	95	75	120	95	125	105	88	64
H. dematioides	85	75	115	105	125	130	100	90
S.phycomyces	77	65	104	112	112	108	116	88

* Growth less than 90 percent of that on the control plates is regarded as a clear inhibition.

Incubation periods varied between 10 and 13 days.

Table 6.—Growth on malt agar with different combinations of boric acid and ground thallus of *Hypogymnea physodes.*

Fungus/treatment	Growth in % of the control (without lichen and boric acid) at various boric acid concentrations (% v/w) in the agar medium				
	0	0.05	0.1	0.5	1.0
H. dematioides					
no lichen	100	NT	86	67	33
0.5% lichen	42	39	61	59	NT
S. phycomyces					
no lichen	100	NT	0	0	0
0.5% lichen	81	11	0	0	NT
*A. pullulans**					
no lichen	100	NT	78	78	63
0.5% lichen	36	49	44	44	NT
*C. piceae**					
no lichen	100	NT	0	0	0
0.5% lichen	71	29	0	0	0

NT = not tested

* The malt agar was buffered to pH 6.5 by the use of a citrate/phosphate buffer.

activity against the target organisms. Different antibiotics are probably most often responsible for the activity although sometimes other types of agents have been revealed, including wall degrading enzymes such as Laminarase or Chitinase or siderophores (14). Even fast colonization, which results in removal of nutrients, might contribute to the effects (21) .

Its evident that the effectivity of all such agents is clearly influenced by the growth media either by media effects on the production of the agents or in the case of compounds like siderophores by higher concentrations of certain metals. The results of the present project also indicate that interactions between the bluestain fungi and the potential bioprotectant *Candida kefyr* gave rise to production of inhibitory agents by *Candida* either through changes in the medium or an effect on metabolism.

Although a vast array of organisms are capable of producing inhibitory compounds, the use of the organisms as bioprotectants against wood damaging fungi might not be possible because they do not grow satisfactorily on wood or they do not tolerate varying environmental conditions (32). The active compounds produced by such organisms might still be characterized and used as models for new synthetic fungicides. Thus it is probably not possible to use the active agents produced by lichens in any other way. In the near future it might be possible to improve bioprotective organisms by the use of modern genetic and cell biology methods.

The use of living bioprotectants for sapstain control might be somewhat doubtful due to possible health effects dependent on heavy spore production or metabolites, notably volatile organic compounds, produced by the organisms which might give allergy or SBS symptoms (11).

Melanin Formation Inhibitors

The importance of melanins for survival and longevity of fungal propagules is well documented (3,13). It was therefore reasonable to assume that a lowering of the melanin content of bluestain fungal hyphae would increase the sensitivity of bluestain fungi.

Several compounds have been shown to repress the melanin synthesis in fungi at concentrations which do not alter the growth rate. Studies have shown that several of these compounds inhibit the enzymatic reduction of two hydroxynaphtalene compounds to scytalone and vermelone, intermediates in the melanin production from 1,8-dihydroxinaphthalene, DHN (3)

A. pullulans required higher doses of the tested melanin inhibitors for melanin formation inhibition than has been reported previously for other fungi. At doses which earlier had been shown to completely inhibit the melanin formation the response varied from growth inhibition to growth stimulation. The treatment could even result in an increased concentration of melanin in the hyphae as judged from chemical analysis of the melanin content.

The combined effects of melanin synthesis inhibitors and conventional fungicides might decrease the required dose of the fungicides. It has also been shown that the effects of a bioprotectant might be stronger against unmelanized hyphae (16).

Electricity

Experiments on the effects of electricity on bluestain fungi (10) revealed that a potential gradient of 1 V/cm corresponding to a current of 15 mA (DC), applied without interruption during at a 2-week experimental period, leads to an inhibition of the growth of these fungi and residual inhibitory effects after the treatment. A potential gradient of 0.5 V/cm also inhibited the bluestain fungi but without residual effects after the treatment. Experiments also revealed that a potential gradient of 10-25 V/cm applied for 30 seconds, 3 times every 24 hours also inhibited the growth of *Aureobasidium pullulans*. The mechanism by which electricity exerts its growth inhibiting effect on bluestain fungi is somewhat unclear, although residual effects after the electrical treatment indicate deprivation of some nutrients or toxic effects of electrolytically produced compounds as contributing factors.

Effect of pH Control

High substrate pH is generally less favorable for growth of fungi. However, as was indicated by previous laboratory tests (30), the field tests confirmed that protection of wood by stabilization of high pH at the surface could not be achieved for longer time periods. This probably depends on a decreased pH with time due to neutralization by CO_2 in the air, movement of acids to the wood surface, or formation of acidic groups at the surface as a result of effects of UV light (23).

Other Possible Methods

Although melanin is a very stable compound, methods for removal of melanin from an already established attack by bluestain fungi might be achieved with the aid of melanin degrading organisms or enzymes isolated from such fungi (26). Previous studies have shown that white rot basidiomycetes are able to colonize and decolorize bluestain fungi in culture or in wood (7,15). However, a potential problem with methods which are directed toward melanin inhibitors or melanin degradation after formation in the wood is other effects of bluestain fungal attack such as increased permeability or decreased decay resistance which were mentioned above.

Conclusions and Possibilities for Non-Conventional Treatments in the Future

Among the tested bioprotectans the inhibitory effects of *S. griseoviridis* and the lichens are especially interesting and should be further studied. The potential use of electricity in certain applications should also be additionally evaluated.

Acknowledgements

The authors are grateful for a grant from the European Council, Air project 1059, associated task 4.4 (Novel treatments). We also want to thank Mr. Staffan Werner, who has carried out much of the laboratory work in this project.

Literature Cited

1. Behrendt, C.J., R.A. Blanchette, and R.L. Farrell. 1995. Biological control of blue-stain fungi in wood. Phytopathology 85:92-97.
2. Behrendt, C.J., R.A. Blanchette, and R.L. Farrell. 1995. An integrated approach using biological and chemical control to prevent blue stain in pine logs. Can. J. Bot. 73:613-619.
3. Bell, A.A. and M.H. Wheeler. 1986. Biosynthesis and functions of fungal melanins. Ann. Rev. Phytopathol. 24:411-451.
4. Benko, R. 1987. Antagonistic effect of some mycorrhizal fungi as biological control of blue-stain. Int. Res. Group on Wood Pres. Document No. IRG/WP/1314.
5. Benko, R. 1988. Bacteria as possible organisms for biological control of blue stain. Int. Res. Group on Wood Pres. Document No. IRG/WP/1339.
6. Benko, R. 1989. Biological control of blue-stain on wood with *Pseudomonas cepacia* 6253: laboratory and field test. Int. Res. Group on Wood Pres. Document No. IRG/WP/1380.
7. Benko, R. and B. Henningsson. 1986. Mycoparasitism by some white rot fungi on blue stain fungi in culture. Int. Res. Group on Wood Pres. Document No. IRG/WP/1304.
8. Bjurman, J. 1988. The importance of blue stain attack for the colonization by wood rotting fungi of wood not in contact with the ground. The International Research Group on Wood Preservation Doc. No. IRG/WP/1349.
9. Bjurman, J. 1992. The protective effect of 23 paint systems on wood against attack by decay fungi-A laboratory study. Holz als Roh und Werkstoff 50:201-206.
10. Bjurman, J. 1996. Growth inhibitory effects of blue-stain fungi of applied electricity fields. The International Research Group on Wood Preservation Doc. No. IRG/WP 96-10167.
11. Bjurman J. and J. Kristensson. 1992. Volatile production by *Aspergillus versicolor* as a possible cause of odor in houses affected by fungi. Mycopathologia 118:173-178.
12. Blanchette, R.A., R.L. Farrell, T.A. Burnes, P.A. Wendler, W. Zimmerman, T.S. Brush, and R.A. Snyder. 1992. Biological control of pitch in pulp and paper production by *Ophiostoma piliferum*. Tappi J. 75:102-106.
13. Bloomfield, B.J. and M. Alexander. 1967. Melanins and resistance of fungi to lysis. J. Bacteriology 93:1,276-1,280.
14. Bruce, A., U. Srinivasdan, H.J. Staines, and T.L. Highley. 1995. Chitinase and laminarinase production in liquid culture by *Trichoderma* spp. and their role in biocontrol of wood decay fungi. International Biodeterioration 337-353.
15. Croan, S.C. and T.L. Highley. 1991. Biological control of the blue stain fungus *Ceratocystis coerulescens* with fungal antagonists. Mat. und Org. 25(4):255-266.
16. De Cal, A. and P. Melgarejo. 1994. Effects of Penicillium frequentans and its antibiotics on unmelanized hyphae of Monilia laxa. Phytopathology 84:1,010-1,014.
17. Farrell, R.L., R.A. Blanchette, T.S. Brush, Y. Hadar, S. Iverson, K. Krisa, P.A. Wendler, and W. Zimmermann. 1993. Cartapip: A biopulping product for control of pitch and resin acid problems in pulp mills. J. Biotechnol. 30:115-122.
18. Henningsson, B. and H. Lundström. 1970. The influence of lichens, lichen extracts and usnic acid on wood destroying fungi. Mat. und Org. 5:19-31.
19. Henningsson, B., H. Lundström, and J. Bjurman. 1988. Stopp för blånad, mögel och röta. Skogsfakta konferens nr 13, 1989. Sveriges Lantbruksuniversitet, Uppsala.
20. Huang, Y., B.L. Wild, and S.C. Morris. 1992. Postharvest biological control of *Penicillium digitatum* decay on citrus fruit by *Bacillus pumilus*. Ann-Appl. Biol. 120:367-372
21. Hulme, M.A. and J.K. Shields. 1970. Biological control of decay fungi in wood by competition for non-structural carbohydrates. Nature 227:300-301.
22. Janisiewicz, W.J. 1988. Biocontrol of postharvest diseases of apples with antagonist mixtures. Phytopatholgy 63:473-473.
23. Kalnins, M.A., C. Steelinck, and H. Tarkow. 1966. Light-induced free radicals in wood. USDA For. Ser. Res. Paper FPL 58, Madison, WI.
24. Klingström, A.E and S.M. Johansson. 1973. Antagonism of Scytalidium isolates against decay fungi. Phytopathology 63, 473-479.
25. Kreber, B. and J.J. Morrell. 1993. Ability of selected bacterial and fungal bioprotectants to limit fungal stain in ponderosa pine sapwood. Wood Fiber Sci. 25:23-24.
26. Liu, Y.T. and Y.Y. Liao. 1995. Isolation of a melanolytic fungus and its hydrolytic activity on melanin. Mycologia. 87(5):651-654.
27. Lundström, H. and B. Henningsson. 1973. The effect of ten lichens on the growth of wood-destroying fungi. Mat. und Org. 8(3):233-246.
28. McLaughlin, R.J., C.L. Wilson, C.P. Chalutz, W.F. Kurtzman, and S.F. Osman. 1990. Characterization and reclassification of yeasts used for biological control of postharvest diseases of fruits and vegetables. Appl. Environ. Microbiol. 56:3,583-3,586.
29. McLaughlin, R.J., C.L. Wilson, S. Droby, R. Ben-Arie, and E. Chalutz. 1992. Biological control of postharvest diseases of grape, peach, and apple with the yeasts *Kloeckera apiculata* and *Candida guillermondi*. Plant. Dis. 76:470-473.
30. Nussbaum, R. and J. Bjurman 1991. Mögel och blånadsskydd genom förändring av träytans pH. Report 9102002 I. ISSN 0283-4634. Träteknikcentrum, Stockholm, Sweden.

31. Potjewijd, R., M.O. Nisperos, J.K. Burns, M. Parish, and E.A. Baldwin. Cellulose-based coatings as carriers for *Candida guillermondi* and *Debaryomyces* sp. in reducing decay of organges. Hortscience 30(7):1,417-1,424.
32. Seifert, K.A., C. Breuil, L. Rossignol, M. Best, and J.N. Saddler. 1988. Screening for microorganisms with the potential for biological control of sapstain on unseasoned lumber. Mat. und Org. 23:81-95.
33. Smilanick, J.L. and R. Denis-Arrue. 1992. Control of green mold of lemons with Pseudomonas species. Plant Dis. 76:481-485.
34. Wilson, C.L. and Chalutz. 1989. Postharvest biological control of Penicillium rots of citrus with antagonistic yeasts and bacteria. Sci. Hort. 40:105-112.

Comparison of Biocides Under Laboratory and Field Conditions

M. H. Freeman, A. D. Accampo, and T. L. Woods

Abstract

The degradation of freshly sawn or unseasoned wood by fungal organisms, specifically sapstain and mold fungi, has been a problem for as long as lumber has been manufactured. The evaluation of fungicides used to protect unseasoned wood has historically been performed in petri dishes, on wood or cellulose wafers, in laboratory chambers, in simulated field tests, and with large scale field evaluations. The purpose of this study was not to compare specific biocide formulation efficacy, but rather to determine the influence on performance results of a controlled laboratory test environment to an accelerated field exposure site. Data resulting from this series of tests indicate that the use of a laboratory chamber without artificial inoculation can closely predict the performance of various biocides when compared to accelerated field conditions, although variability between methods remains high.

Introduction

The degradation of freshly sawn or unseasoned wood by fungal organisms, specifically sapstain and mold fungi, has been a problem for as long as lumber has been manufactured. The evaluation of fungicides used to protect unseasoned wood has historically been performed in petri dishes, on wood or cellulose wafers, in laboratory chambers, in simulated field tests, and with large scale field evaluations.

In this study, two specific test methods, using similar experimental design, at two separate test locations were used to evaluate the overall performance of 40 biocide combinations in a comparison of a laboratory exposure and an accelerated field test exposure.

Many scientists and researchers (3-7,11,12,15-17) have performed fungicidal effectiveness and efficacy tests on wood in attempts to screen different biocides for potential commercialization as wood treatments. Additionally, these and other researchers (5,7,9-12) have also screened fungicides in field trials or in accelerated field test methods. Currently, the American Wood Preservers' Association is attempting to categorize the different test methods in a compendium and standardize a universal test method which will ultimately be used to standardize testing, similar to the method subscribed to by the American Society of Testing and Materials (1).

It is currently estimated that over 36 million m^3 of wood is treated with anti-sapstain preservative chemicals on an annual basis. This amount of chemicals, including the chlorophenates, could possibly exceed 5,000 tons worldwide on an annual basis. Although small in comparison to the total pesticide market, it does represent significant sales of biocides and results in the protection of over 12 billion board feet of commercially marketable lumber and timber. Thus, to effectively develop models which can be used to rapidly screen formulations under laboratory and accelerated field conditions, is a worthy task.

In this study, an accelerated field test site located in a warm, moist environment is compared to a controlled temperature, controlled humidity test chamber on the same species, and boards with similar width and thickness.

Materials and Methods

In this study, the freshly sawn wooden substrate used was Southern yellow pine (*Pinus* spp.) sapwood. No specific effort was attempted to classify the exact southern pine sub-category (shortleaf, slash, etc.), however the use

M. H. Freeman, Technical Manager, A. D. Accampo, Research Chemist, and T. L. Woods, Division Technical Manager, ISK Biosciences Corp., Industrial Biocides Division, Memphis, Tennessee.

of heartwood-sapwood indicator as well as visual sorting, allowed the boards evaluated to contain >95 percent sapwood. The test boards were obtained rough-planed from a central Mississippi sawmill and were cut to lengths of 0.5 meter for laboratory evaluation and 1.0 meter for field evaluation. Lumber thickness was approximately 28 cm.

The time elapsed from sawing the boards from pine logs until treatment was less than 48 hours. Treatment was performed by immersing the wooden test pieces in the appropriate biocide formulation for 30 seconds in an inert (polyester resin coated dip vat). All biocides tested were diluted with City of Memphis tap water (pH = 6.85, water hardness <320 ppm total Ca/Mg) to a concentration of 1 percent (w/w) or approximately 1 part of biocide formulation to 99 parts of water. Where additional chemicals or treatments were added to the biocide formulation, an attempt was made to test these at commercial use rates if the additive was a commercial product, or at a rate recommended by the manufacturer.

In these tests, a total of eight biocides were tested in a total of 40 total formulations. A list of the biocides which were known to be in some of the formulations and the common name and abbreviation for the biocide is located in Table 1.

A complete listing of the 40 biocide formulations used to compare the two test sites/test methodologies is located in Table 2. It should be noted here that wherever practicable, the use of common names and abbreviations for the products in test are used. There is no attempt to cover all registered trademarks, proprietary product names, or identifying symbols. Wherever possible, the product is listed by it's active ingredient make-up versus it's commercial or trademarked name. Certain proprietary additives are shown with their commercial name or the name furnished by the vendor or supplier of these chemicals. As is common in industry, the exact composition of these additives is not disclosed on the Material Safety Data Sheets or product labels, and the product is refereed to in this study by it's common name. We apologize if we have left out any ®, ™ , or any trademarked logos or names in this paper. It was not done intentionally or with malice to any product vendor or supplier.

Table 1.—List of known active ingredients evaluated and their common and chemical names.

Abbreviation	Common name	Chemical name
CTL	Chlorothalonil	Tetrachloroisophthalonitrile
CARB	Carbendazim	2-(methoxycarbonylamino)-benzimidazole
DDAC	Quat	didecylydimethyl ammonium chloride
IPBC	Polyphase	3-iodo, 2-propynyl butyl carbamate
ITA	Kathon	isothiazilone
MBT	Methylene bis	methylene bis-thiocyanate
Cu8Q	Copper-8-Quinolinolate	Copper-8-hydroxyquinolate
TCMTB	TCMTB	2-(thiocyanomethylthio)benzothiazole

Following treatment of the sample boards, the short boards (0.5 m) were placed into a controlled environment chamber, dead stacked (without stickers) six boards high in 40 separate treated wood rows. The temperature of the chamber was maintained at >21°C and <32°C. The relative humidity of the chamber was maintained at >70 percent relative humidity by the periodic injection of both water vapor and/or steam.

Following treatment the larger sample boards (1.0 m) were placed into a commercial truck van body after being shrink-wrapped (to maintain moisture until field exposure) and transported to Miami, Florida, where they were then dead stacked (without stickers) ten boards high in 40 separate treated wood rows. The treated wood was supported on untreated wooden stringers supported approximately 0.2 m off the ground using commercial hollow concrete cinder blocks. The stacks of the treated wood were then loosely covered with translucent, polyethylene sheeting, of approximately 20-mil plastic thickness. The exact geographical location of the test site is 25° 48″ north latitude, 80° 12″ west longitude near Miami, Florida. The average daily temperature during the test period was 28°C, and the average relative humidity was 83 percent. This outdoor site has constant monitoring for temperature, relative humidity, and sun angle since it is a commercial coatings evaluation site.

Following exposure, the test boards were separated manually, and all surfaces examined for fungal discoloration (sapstain and mold growth). Each board was individually rated on a scale ranging from 0 to 100, where 0 = no control, and 100 = perfect, no fungal discoloration or stain. The total of all the boards at each test site were evaluated at 3 weeks and 6 weeks of exposure. An additional 10 week exposure period of evaluation was performed on the boards in the humidity chamber in Memphis, Tennessee, due to ease of data acquisition. All numerical values were transformed to a 0 to 10 scale by dividing the mean value by 10 for ease of data analysis. Using the 0 to 10 scale, a 0 indicates no control of stain/mold, and a 10 indicates 100 percent control of stain and mold. The individual ratings will be presented later in the paper.

Treatment Number	Identification Code	Active Ingredient(s) and Percentage (IF KNOWN)	Additive(1) and Rate of Use(W/W)	Additive(2) and Rate of Use(W/W)
1	SBP	IPBC-6%		
2	SBP	IPBC-6%	ISK Brite @ 0.25 %	
3	SBP	IPBC-6%	ISK-1 @ 0.2 %	
4	SBP	IPBC-6%	ISK-1 @ 0.125 %	
5	NP	IPBC-7.6%, DDAC-64.8%	ISCA 250 @ 0.25 %	
6	NPP	IPBC-7.55%, DDAC + ITA		
7	NP	IPBC-7.6%, DDAC-64.8%	Iron FixT @ 0.25 %	
8	NPP	IPBC-7.55%, DDAC + ITA	Iron FixT @ 0.25 %	
9	PQ	CU8Q-5.4%	Borax @ 1.5 %	CTL
10	PQ	CU8Q-5.4%	Borax @ 1.5 %	PROPRIETARY
11	BU9	TCMTB-10%, MBT-10%	Busperse 293 @ 0.25 %	
12	BU18	TCMTB-30%	Busperse 293 @ 0.25 %	
13	TBC	CTL-36%, CARB-8%	Borax @ 1.5 %	
14	TBC	CTL-36%, CARB-8%	ISK Brite @ 0.5 %	
15	TBC	CTL-36%, CARB-8%	ISK Brite @ 0.25 %	
16	TBC	CTL-36%, CARB-8%	ISK Brite @ 0.125 %	
17	TBC	CTL-36%, CARB-8%	Borax @ 0.75 %	
18	TBC	CTL-36%, CARB-8%		
19	HE	CU8Q , CARB		
20	NX	CTL-15%, MBT-15%		
21	NXHC	CTL, MBT	ISK Brite @ 0.25 %	
22	NX	CTL-15%, MBT-15%	ISK Brite @ 0.25 %	
23	NX	CTL-15%, MBT-15%	ISK-1 @ 0.125 %	
24	NXP	CTL-15%, MBT-15%		
25	NX40P	CTL, MBT		
26	NX	CTL-15%, MBT-15%	ISK-1 @ 0.125 %	PROPRIETARY
27	NX	CTL-15%, MBT-15%	ISK-1 @ 0.125 %	PROPRIETARY
28	TB	CTL-40.4%	T13 P @ 0.5 %	
29	SBP250	PROPRIETARY MIXTURE		
30	SBP250	PROPRIETARY MIXTURE	ISK-1 @ 0.125 %	
31	SBP252	PROPRIETARY MIXTURE		
32	SBP257	PROPRIETARY MIXTURE	ISK-1 @ 0.125 %	
33	SBP258	PROPRIETARY MIXTURE	ISK-1 @ 0.125 %	
34	SBP262	PROPRIETARY MIXTURE	ISK-1 @ 0.125 %	
35	PQC	PROPRIETARY MIXTURE		
36	CTLT	PROPRIETARY MIXTURE		
37	NNX2	PROPRIETARY MIXTURE		
38	NNX3	PROPRIETARY MIXTURE		
39	NNX4	PROPRIETARY MIXTURE		
40	Water Control	WATER		

Table 2.—List of treatments in this test series.

Results and Conclusions

The individual board ratings for the 3-week and 6-week readouts can be located in Tables 3 through 7. Table 8, a Summary Table, is a compilation of all the mean values from both test sites at all three time periods. A brief overview of the data indicates that a favorable comparison can be seen when observing a specific biocide performance at either location. Moreover, a cursory review of the overall data seems to indicate very good prediction models may be derived from laboratory data as compared to the accelerated field model. No data exists to compare how much the field model is accelerated over commercial practice, but it should be noted that typical values found in the field model of 6.0 are considered commercially viable products when used under normal commercial conditions.

Table 9 summarizes the results from the Tennessee and Florida exposure sites. As shown in the table the overall averages, averaged over the 39 treatments (the Control was excluded), at the two sites were very similar at both 3 and 6 weeks.

Table 10 shows the corresponding standard deviations. The first set of standard deviations are the *between-treatment* standard deviations. That is, each is the standard deviation of the 39 treatment averages (averages over the six boards in Tennessee and ten in Florida) for a particular site and week. The second set of standard deviations are the pooled averages, over the treatments, of the 39 between-board standard deviations.

From Table 10 we note that there is little difference in variation between the Tennessee and Florida sites at week 3; at week 6, the variation is slightly greater at the Florida

Treatment Number	Replicate Number 1	Replicate Number 2	Replicate Number 3	Replicate Number 4	Replicate Number 5	Replicate Number 6	Mean
1	50	55	60	50	60	65	5.7
2	55	75	50	60	100	90	7.2
3	55	65	75	80	90	90	7.6
4	90	100	90	95	95	95	9.4
5	85	80	90	75	90	80	8.3
6	90	95	85	80	85	90	8.8
7	95	85	85	75	90	75	8.4
8	90	95	85	90	90	75	8.8
9	95	100	100	100	100	100	9.9
10	80	85	95	95	95	95	9.1
11	60	80	80	60	90	85	7.6
12	100	100	100	95	95	90	9.7
13	95	95	90	75	100	100	9.3
14	90	100	95	100	95	100	9.7
15	80	100	95	90	95	100	9.3
16	95	100	100	95	85	95	9.5
17	90	85	100	95	80	95	9.1
18	100	95	90	100	95	100	9.7
19	95	100	95	95	95	95	9.6
20	100	95	95	95	95	100	9.7
21	95	95	100	100	100	100	9.8
22	100	100	100	100	100	100	10.0
23	100	100	100	100	95	100	9.9
24	95	100	100	100	100	100	9.9
25	100	100	95	100	100	100	9.9
26	100	95	100	100	100	100	9.9
27	100	100	100	100	100	100	10.0
28	95	95	85	95	95	90	9.3
29	80	90	80	60	65	75	7.5
30	95	95	85	90	55	90	8.5
31	95	90	85	90	75	95	8.8
32	90	90	75	95	75	90	8.6
33	95	75	85	95	75	90	8.6
34	95	95	90	95	90	95	9.3
35	95	95	95	100	95	100	9.7
36	90	90	95	85	90	95	9.1
37	90	95	90	95	85	90	9.1
38	90	95	90	95	90	95	9.3
39	100	95	95	95	90	100	9.6
40	0	5	0	0	0	5	0.2

Table 3.—Tennessee 3-week readout, June 3, 1996.

Treatment Number	Replicate Number 1	Replicate Number 2	Replicate Number 3	Replicate Number 4	Replicate Number 5	Replicate Number 6	Replicate Number 7	Replicate Number 8	Replicate Number 9	Replicate Number 10	Mean
1	95	95	95	85	95	95	90	70	70	100	8.9
2	90	70	80	90	95	95	9	90	95	85	8.0
3	90	85	90	95	80	100	95	95	95	95	9.2
4	90	90	90	80	70	90	90	95	75	80	8.5
5	90	85	75	80	85	80	75	70	80	90	8.1
6	85	80	80	80	75	80	90	95	80	75	8.2
7	75	80	80	90	90	80	90	95	60	90	8.3
8	85	80	75	80	90	90	85	80	80	90	8.4
9	95	95	95	95	100	100	100	95	95	95	9.7
10	95	95	90	95	90	95	95	95	90	95	9.4
11	95	100	100	100	100	95	95	90	95	95	9.7
12	95	100	95	100	90	80	95	95	95	95	9.4
13	100	100	100	100	95	95	95	100	95	100	9.8
14	100	100	100	100	100	100	100	95	100	100	10.0
15	100	100	100	100	95	100	100	100	100	100	10.0
16	95	100	100	100	100	95	100	95	100	100	9.9
17	100	100	100	100	95	100	95	95	95	95	9.8
18	95	100	100	95	95	95	100	95	95	100	9.7
19	95	95	100	95	95	95	95	90	85	90	9.4
20	100	100	100	100	100	90	100	100	100	100	9.9
21	95	100	100	100	100	100	95	100	100	100	9.9
22	100	100	100	100	100	100	95	100	90	95	9.8
23	100	100	100	95	100	95	100	100	100	100	9.9
24	100	100	95	100	100	100	100	100	100	100	10.0
25	100	100	100	100	100	100	100	95	100	95	9.9
26	100	100	100	95	100	100	95	100	100	100	9.9
27	95	95	100	95	100	95	85	100	100	100	9.7
28	95	95	100	95	95	95	95	90	100	95	9.6
29	85	80	75	80	85	90	85	65	90	85	8.2
30	20	55	65	70	75	85	85	90	80	85	7.1
31	60	60	95	85	70	75	80	90	95	95	8.1
32	25	35	50	60	90	90	85	70	75	80	6.6
33	95	85	75	90	95	80	75	90	85	95	8.7
34	85	95	90	95	100	95	100	95	85	95	9.4
35	90	100	100	95	95	90	85	90	90	90	9.3
36	95	100	100	95	95	95	100	95	90	95	9.6
37	75	90	100	100	95	90	95	95	90	90	9.2
38	90	100	95	95	100	90	95	90	85	95	9.4
39	100	95	100	100	100	100	90	85	75	100	9.5
40	5	0	5	0	5	0	5	10	0	0	0.3

Table 4.—Florida 3-week readout, June 7, 1996.

Treatment Number	Replicate Number 1	Replicate Number 2	Replicate Number 3	Replicate Number 4	Replicate Number 5	Replicate Number 6	Mean
1	25	30	40	35	45	75	**4.2**
2	40	60	50	40	90	75	**5.9**
3	50	30	40	50	55	55	**4.7**
4	75	100	80	95	90	90	**8.8**
5	60	60	70	60	65	55	**6.2**
6	85	90	80	90	75	80	**8.3**
7	85	70	80	80	85	85	**8.1**
8	80	90	80	95	80	80	**8.4**
9	95	95	100	100	100	100	**9.8**
10	60	60	75	90	90	85	**7.7**
11	40	55	55	25	85	80	**5.7**
12	90	95	95	95	85	90	**9.2**
13	60	70	25	40	90	80	**6.1**
14	65	100	90	95	95	95	**9.0**
15	70	90	95	85	95	100	**8.9**
16	95	100	100	85	75	80	**8.9**
17	90	90	100	100	80	95	**9.3**
18	95	95	65	90	90	95	**8.8**
19	95	100	90	85	80	95	**9.1**
20	95	95	95	95	95	100	**9.6**
21	95	95	100	100	100	100	**9.8**
22	100	100	100	100	95	100	**9.9**
23	95	100	95	95	95	95	**9.6**
24	95	100	95	95	100	95	**9.7**
25	95	90	100	100	100	100	**9.8**
26	95	90	95	95	100	95	**9.5**
27	100	100	100	100	95	95	**9.8**
28	90	85	80	90	95	90	**8.8**
29	65	70	65	45	50	65	**6.0**
30	90	90	75	85	55	85	**8.0**
31	85	60	30	80	60	85	**6.7**
32	95	80	60	80	50	55	**7.0**
33	70	55	50	75	60	90	**6.7**
34	85	70	60	70	70	65	**7.0**
35	90	80	90	100	85	85	**8.8**
36	75	75	80	65	50	90	**7.3**
37	80	90	60	75	65	75	**7.4**
38	80	85	80	95	75	75	**8.2**
39	75	95	90	90	75	90	**8.6**
40	0	0	0	0	0	5	**0.1**

Table 5.—Tennessee 6-week readout, June 21, 1996.

Treatment Number	Replicate Number 1	Replicate Number 2	Replicate Number 3	Replicate Number 4	Replicate Number 5	Replicate Number 6	Replicate Number 7	Replicate Number 8	Replicate Number 9	Replicate Number 10	Mean
1	75	60	40	50	50	60	50	55	85	75	**6.0**
2	40	85	90	85	80	75	75	50	40	35	**6.6**
3	85	95	90	85	90	80	50	40	90	65	**7.7**
4	10	35	65	75	75	15	25	50	45	45	**4.4**
5	25	15	10	65	75	85	75	40	65	55	**5.1**
6	65	55	75	85	65	55	25	15	20	10	**4.7**
7	60	70	90	90	85	75	75	45	10	10	**6.1**
8	80	20	60	55	75	65	50	30	10	20	**4.7**
9	45	80	85	100	100	100	95	95	90	75	**8.7**
10	90	85	90	90	80	75	90	70	75	75	**8.2**
11	80	90	95	95	100	95	85	80	95	100	**9.2**
12	95	95	95	90	75	50	80	95	90	95	**8.6**
13	100	95	100	95	90	90	100	95	95	90	**9.5**
14	100	100	95	95	95	95	95	100	100	95	**9.7**
15	100	100	100	100	100	95	100	90	85	100	**9.7**
16	95	100	100	95	90	100	95	95	100	90	**9.6**
17	95	90	95	95	95	100	100	95	95	95	**9.6**
18	95	90	85	95	95	90	95	95	85	80	**9.1**
19	75	85	90	95	100	80	50	75	65	90	**8.1**
20	90	100	95	95	75	65	90	95	100	100	**9.1**
21	100	100	95	90	100	100	90	100	100	100	**9.8**
22	20	75	100	95	95	100	100	95	95	90	**8.7**
23	95	100	95	100	95	95	95	95	95	100	**9.7**
24	95	100	100	100	100	95	95	95	95	100	**9.8**
25	90	100	95	100	100	100	100	95	90	95	**9.7**
26	95	100	100	95	100	95	95	95	100	95	**9.7**
27	100	100	100	75	85	75	100	100	100	95	**9.3**
28	100	95	95	100	100	95	100	100	90	95	**9.7**
29	45	75	85	45	50	55	60	20	20	20	**4.8**
30	60	50	75	75	65	45	50	30	10	5	**4.7**
31	65	80	90	80	65	35	50	75	10	5	**5.6**
32	50	25	55	60	75	85	80	20	15	5	**4.7**
33	50	50	75	75	85	80	85	15	10	65	**5.9**
34	80	70	80	100	95	100	95	50	75	40	**7.9**
35	90	90	95	90	90	90	90	100	100	90	**9.3**
36	95	95	100	90	95	95	100	100	100	100	**9.7**
37	95	95	100	95	85	95	95	100	80	65	**9.1**
38	95	90	95	95	90	100	95	90	100	70	**9.2**
39	100	95	95	100	100	100	100	95	100	100	**9.9**
40	0	0	0	0	0	0	0	0	0	0	**0.0**

Table 6.—Florida 6-week readout, June 28, 1996.

Treatment Number	Replicate Number 1	Replicate Number 2	Replicate Number 3	Replicate Number 4	Replicate Number 5	Replicate Number 6	Mean
1	70	50	20	45	50	0	**3.9**
2	75	95	35	15	35	20	**4.6**
3	65	15	15	40	35	15	**3.1**
4	55	75	65	50	75	65	**6.4**
5	65	70	80	55	60	60	**6.5**
6	90	95	95	90	80	80	**8.8**
7	75	75	65	80	65	65	**7.1**
8	55	70	95	80	90	60	**7.5**
9	85	100	95	100	95	100	**9.6**
10	85	90	90	85	75	60	**8.1**
11	80	25	10	50	50	15	**3.8**
12	80	80	90	90	100	60	**8.3**
13	95	80	55	45	60	45	**6.3**
14	85	90	95	95	95	60	**8.7**
15	90	95	85	90	90	75	**8.8**
16	75	70	60	85	95	90	**7.9**
17	85	85	95	75	85	95	**8.7**
18	85	80	90	60	80	90	**8.1**
19	95	75	80	90	90	80	**8.5**
20	85	80	80	85	85	90	**8.4**
21	95	85	90	85	95	95	**9.1**
22	95	90	95	95	90	85	**9.2**
23	95	90	95	95	85	80	**9.0**
24	100	95	95	95	95	80	**9.3**
25	90	90	95	90	90	95	**9.2**
26	95	90	90	90	90	85	**9.0**
27	90	95	95	95	95	95	**9.4**
28	85	95	95	90	90	95	**9.2**
29	70	70	50	65	60	55	**6.2**
30	80	60	65	50	65	75	**6.6**
31	80	50	70	50	55	60	**6.1**
32	50	35	60	45	60	90	**5.7**
33	85	40	50	30	45	70	**5.3**
34	70	55	45	20	0	80	**4.5**
35	90	85	90	95	65	90	**8.6**
36	90	75	65	80	75	70	**7.6**
37	85	80	85	65	90	75	**8.0**
38	70	75	95	70	85	75	**7.8**
39	90	80	90	90	90	90	**8.8**
40	0	0	0	0	0	0	**0.0**

Table 7.—Tennessee 10-week readout, July 19, 1996.

TREATMENT NO.	FL-3 wk	FL - 6 wk	TN - 3 wk	TN - 6 wk	TN -10 wk
1	8.9	6.0	5.7	4.2	3.9
2	8.8	6.6	7.2	5.9	4.6
3	9.2	7.7	7.6	4.7	3.1
4	8.5	4.4	9.4	8.8	6.4
5	8.1	5.1	8.3	6.2	6.5
6	8.2	4.7	8.8	8.3	8.8
7	8.3	6.1	8.4	8.1	7.1
8	8.4	4.7	8.8	8.4	7.5
9	9.7	8.7	9.9	9.8	9.6
10	9.4	8.2	9.1	7.7	8.1
11	9.7	9.2	7.6	5.7	3.8
12	9.4	8.6	9.7	9.2	8.3
13	9.8	9.5	9.3	6.1	6.3
14	10.0	9.7	9.7	9.0	8.7
15	10.0	9.7	9.3	8.9	8.8
16	9.9	9.6	9.5	8.9	7.9
17	9.8	9.6	9.1	9.3	8.7
18	9.7	9.1	9.7	8.8	8.1
19	9.4	8.1	9.6	9.1	8.5
20	9.9	9.1	9.7	9.6	8.4
21	9.9	9.8	9.8	9.8	9.1
22	9.8	8.7	10.0	9.9	9.2
23	9.9	9.7	9.9	9.6	9.0
24	10.0	9.8	9.9	9.7	9.3
25	9.9	9.7	9.9	9.8	9.2
26	9.9	9.7	9.9	9.5	9.0
27	9.7	9.3	10.0	9.8	9.4
28	9.6	9.7	9.3	8.8	9.2
29	8.2	4.8	7.5	6.0	6.2
30	7.1	4.7	8.5	8.0	6.6
31	8.1	5.6	8.8	6.7	6.1
32	6.6	4.7	8.6	7.0	5.7
33	8.7	5.9	8.6	6.7	5.3
34	9.4	7.9	9.3	7.0	4.5
35	9.3	9.3	9.7	8.8	8.6
36	9.6	9.7	9.1	7.3	7.6
37	9.2	9.1	9.1	7.4	8.0
38	9.4	9.2	9.3	8.2	7.8
39	9.5	9.9	9.6	8.6	8.8
Water Control	0.3	0.0	0.2	0.1	0.0

Table 8.—Summary data: both test sites by treatment number.

Table 9.—Average efficiacy at the exposure sites.

Week	Tennessee	Florida	Difference
3	9.05	9.18	–0.13
6	8.08	8	0.09
10	7.48	N/A	N/A

Table 10.—Average variation at the exposure sites.

	Between treatments			Between boards	
Week	TN	FL	Diff.	TN	FL
3	0.9	0.8	0.9	0.7	0.7
6	1.5	1.9	1.8	1.1	1.7
10	1.8	N/A	N/A	1.4	N/A

Table 11.—Treatment class means.

Treatment class (number in class)	Class means by site		Overall	
	FL @ 6	TN @ 10	Mean	Std. Dev.[a]
NX (8)	9.5	9.1	9.3	0.8
PQC (1)	9.3	8.6	8.9	0.8
TB (7)	9.6	8.2	8.9	0.9
NNX (3)	9.4	8.2	8.8	0.8
PQ (2)	8.5	8.8	8.6	1.1
CTL T (1)	9.7	7.6	8.6	0.7
HE (1)	8.1	8.5	8.3	1.2
BU (2)	8.9	6.1	7.5	1.7
NP (4)	5.2	7.5	6.3	2.1
SBP (10)	5.8	5.2	5.5	2.2
Average	8.0	7.5	7.7	1.5

[a] The between-board standard deviation pooled over the treatments within the class.

site, both between the treatments and between the boards for a given treatment. However, we do not consider that the larger between-treatment variation at the Florida site indicates better treatment separation. It simply reflects the overall greater variation that would be expected in a "natural" versus a "controlled" environment. (Note: the difference in the number of boards per treatment at the sites makes little difference in the comparison of the between-treatment estimates of variation and no difference in the comparison of the between-board estimates.)

The closeness of the Tennessee and Florida means in Table 9 suggests good agreement between the measurement at the two sites. However, for some treatments the differences in the averages were quite large.

Several attempts to model the Florida results as a function of the Tennessee results were unsuccessful—the fits were too poor. The correlations between the 39 weekly treatment means within each site were high (0.8 to 0.9); this was expected, since these are the same boards, just inspected at different times. Also expected, however, the correlations between the Florida and Tennessee results were lower (0.5 for both week 3 and week 6); too low to yield a reasonable prediction model. In each case, it was the treatments with the lower averages that were so variable that they could not be fit well.

Comparison of Treatments

For the comparison of the 39 treatments, the 10-week Tennessee means and the 6-week Florida means were averaged. Also, the treatments were divided into 10 classes based on their core products. The class means are shown in Table 3. Tables 3 through 7 and Summary Table 8 show the averages for each of the 39 treatments: in all the Tables, the treatments are listed in the original order: no tables were sorted as to performance rank order. The treatments are later sorted into their respective classes, and the classes are ordered from highest class average to lowest.

An analysis of variance was performed on the 39 treatment means (the averages of the 10-week Tennessee and 6-week Florida means) to assess the differences among the products. There were significant differences among the product classes; the 10 classes explained over 90 percent of the variation among the treatments. No formal comparison of the within-class treatments was performed, but the residual within-class variation was no larger than would be expected based on the between-board variation. No significant differences among the Product 2's were detected.

The NX class had the highest average. In Table 11, those treatments that were not significantly different from NX are included within the same box. Thus, based on the analysis of variance, the only treatments that were significantly worse than NX at the 5 percent level were BU, NP, and SBP. Note, however, that some of the classes were small, containing only one or two treatments.

A final note on the variation: the average *between-board* standard deviations for the classes are shown in Table 11 with the class means. Comparing these standard deviations and means, we can see that as the means gets smaller, the standard deviations get larger. This was reflected indirectly in the differences between the corresponding Tennessee and Florida means. It is suspected that the ratings behave similar to proportions. That is, the rating is on a scale of 0 or 1 to 10; near 0 or near 10 there is relatively little variation in the ratings from board to board, but further away from the scale endpoints, the variation increases. An analysis of the transformed data, using a transformation analogous to the arc-sine, was performed. However, in this case, the results were similar to those obtained for the untransformed data.

This work indicates that although significant progress has been made in correlating small laboratory experiments to those of accelerated field trials, more work needs to be performed in this area. Additionally, more tests need to be performed to correlated small scale tests with those commercially used practices.

Acknowledgments

Special thanks to Dr. John Schollenberger of Ricerca for his assistance in all of the statistical manipulation of the data. Additionally, special thanks to Bob Andrews, and Dr. Jim Shaw of Weyerhaeuser Corp., and Gary Chang, Tim Reid, and Marlin Thyer for their assistance in securing raw materials for the trial.

Literature Cited

1. American Society for Testing and Materials. 1988. Standard method for testing fungicides for controlling sapstain and mold on unseasoned lumber (laboratory method). ASTM Standard D4445-84. Philadelphia, PA.
2. Bravery, A.F. 1978. Chlorophenols in wood treatments. Suppl. to Timber Trade J. (Oct. 21):2-6 and 10.
3. Butcher, J.A. 1973. Laboratory screening trials for new prophylactic chemicals against sapstain and decay in sawn timber. Mat. und Org. 9(1):51-70.
4. Butcher, A.F. 1979. Antisapstains on the N.Z. market. Forest Ind. Rev. 10(4):12-13.
5. Butcher, A.F. and J. Drysdale. 1974. Field trials with captafol — an acceptable antisapstain chemical. Forest Prod. J. 24(11):28-30.
6. Butcher, A.F. and J. Drysdale. 1978. Laboratory screening trials with chemicals for protection of sawn timber against mould, sapstain and decay. Int. Biodeterior. Bull. 14(1):11-19.
7. Cassens, D.L. and W.E. Eslyn. 1983. Field trials of chemicals to control sapstain and mold on yellow poplar and southern yellow pine lumber. Forest Prod. J. 33(10):52-56.
8. Clark, J.E. and R.S. Smith. 1979. Culture collection of wood inhabiting fungi. Forintek Canada Corp. Tech. Rept. 2, Vancouver, B.C.
9. Cserjesi, A.J. 1977. Prevention of stain and mould in lumber and board products. In: Goldstein, I.S. Wood Technology, Chemical Aspects. ASS Symp. Series No. 43, Am. Chem. Soc., Washington, D.C.
10. Cserjesi, A.J. 1980. Field testing fungicides on unseasoned lumber — recommended procedure. Forintek Canada Corp., Vancouver, B.C. Tech. Rept. 16. 12 p.
11. Cserjesi, A.J. and E.L. Johnson. 1982. Mold and sapstain control: laboratory and field tests on 44 fungicidal formulation. Forest Prod. J. 32(10):59-68.
12. Cserjesi, A.J. and J.W. Roff. 1970. Accelerated laboratory test for evaluating the toxicity of fungicides for lumber. Mater. Res. and Stand. 10(3):18-19, 59-60.
13. Cserjesi, A.J. and J.W. Roff. 1975. Toxicity tests of some chemicals against certain wood-staining fungi. Int. Biodeterior. Bull. 11(3):90-96.
14. Davis, E.F., B.L. Tuma, and L.C. Lee. 1959. Handbook of Toxicology; Vol. V, Fungicides. W.B Sounders Co., Philadelphia and London.
15. Dickinson, J.D. 1977. The effective control of blue-stain mould on freshly-felled timber. Holzforschung 31(4):121-125.
16. Drysdale, J.A. and A.F. Preston. 1982. Laboratory screening trials with chemicals for protection of green timber against fungi. New Zealand J. of Forestry Sci. 12(3):457-466.
17. Eslyn, W.E. and D.L. Cassens. 1983. Laboratory evaluation of selected fungicides for control of sapstain and mold on southern pine lumber. Forest Prod. J. 33(4):65-68.
18. Farm Chemicals Handbook. 1980. Meister Publishing Co., Willoughby, OH.
19. Frear, D.E.H. 1969. Pesticide Index (4th ed.). College Science Publishers, State College, PA.
20. Hayward, P.J., W.J. Rae, and J. Duff. 1984. Mixtures of fungicides screened for the control of sapstain on *Pinus radiata*. Inter. Research Group on Wood Preserv., Document No. IRG/WP/3307.
21. Hulme, M.A. and J.F. Thomas. 1979. Control of fungal sapstain with alkaline solutions of quaternary ammonium compounds and with tertiary amine salts. Forest Prod. J. 29(11):26-29.
22. Lindgren, R.M. 1942. Temperature, moisture, and penetration studies of wood-staining *Ceratostomella* in relation to their control. USDA, Washington, D.C. Technical Bulletin 807. 35 p.
23. Merck & Co. Ltd. 1960. The Merck Index of Chemicals and Drugs. Merck & Co. Inc., Rahway, NJ.
24. Micales, J.A., T.L. Highley, and A.L. Richter. 1989. The use of chlorothalonil for protection against mold and sapstain fungi. I. Laboratory evaluation. Inter. Research Group on Wood Preserv., Document No. IRG/WP/3515.
25. Miller, D.J. and J.J. Morrell. 1989. Controlling sapstain: trials of product group I on selected western softwoods. Forest Res. Lab., Oregon State Univ., Corvallis, OR. Res. Bull. 65. 12 p.
26. Presnell, T.L. and D.D. Nicholas. 1990. Evaluation of combinations of low hazard biocides in controlling mold and stain fungi on southern pine. Forest Prod. J. 40(2):57-61.
27. Richardson, B.A. 1972. Sapstain control. Paperi ja puu (10):613-624.
28. Roff, J.W. and A.J. Cserjesi. 1965. Chemical preventatives used against mould and sapstain in unseasoned lumber. B.C. Lumberman 49(5):90-98.
29. Roff, J.W., A.J. Cserjesi, and G.W. Swann. 1980. Prevention of sapstain and mould in packaged lumber. Forintek Canada Corp. Tech. Rept. 14R, Vancouver, B.C.
30. Sapienza, M.S. and M.R. Purvis. 1975. Tetrachloroisophthalonitrile composition for treatment of lumber. Canadian Patent 967,482.

31. Sassaman, J.F., M.M. Jacobs, P.H. Chin, S. Hsia, R.J. Pienta, and J.M. Kelly. 1986. Pesticide background statements Volume II. Fungicides and Fumigants. Agri. Handb. No. 661. USDA Forest Serv., Washington, D.C.
32. Shields, J.K., R.L. Desai, and M.R. Clarke. 1974. Ammoniacal zinc oxide treatment as an inhibitor of fungi in pine lumber. Forest Prod. J. 24(2):54-57.
33. Smith, R.S., A. Byrne, and A.J. Cserjesi. 1987. New fungicidal formulations protect Canadian lumber. Forintek Canada Corporation, Vancouver, B.C. Special Publication Sp-25. 3 p.
34. Thompson, W.T. 1976. Agricultural Chemicals — Book IV, Fungicides. Thompson Publications, Fresno, CA.
35. Torpeson, D.C. 1969. Fungicides. Vol. 2, An Advanced Treatise. Academic Press, New York and London.
36. Unligil, H.H. 1976. Prevention of fungal stain on pine lumber: laboratory screening tests with fungicides. Forest Prod. J. 26(1):32-33.
37. Verrall, A.F. and P.V. Mook. 1951. Research on chemical control of fungi in green lumber, 1940-51. USDA Tech. Bull. 1046, Washington, DC.